LION RAMPANT and WINGED

by

Alan Robertson

A Commemorative History of Scottish Aviation Limited

Predecessor Company of

British Aerospace PLC Civil Aircraft Division Prestwick

First published in Industry Year, 1986.

ISBN 0 9511123 1 7 (Hardback)

ISBN 0 9511123 0 9 (Paperback)

Cover: I. Macleod

Designed by I.Macleod

Printed and bound
by McCorquodale (Scotland) Ltd.

for the publisher, Alan Robertson
4 Citadel Way
Barassie, KA10 6U
Scotland

Contents

A Commemorative History of Scottish Aviation Limited

Acknowledgements

It would not have been possible to write this book without the assistance of a great many people whose contributions have ranged from single anecdotes to extended correspondence, from lengthy interviews to the loan of photographs, documents and other memorabilia. It is not possible to mention them all here but I would like to record my sincere thanks for all this help and in particular to

Ian Andrew, Elizabeth Ashton, Andrew Bartleman, P J Bish, the late Noel Capper, John Chalmers, John Coombes, D J Corley, A J Cormack, Ann Crawford, Jack Dickie, David Dorman, Ron Frost, Jim Guthrie, Don and Chris Hall, John Hope, Tommy Hope, Mervyn Hunter, Jake Hyslop, Mary Kennedy, A Stewart Kennedy, John McGowan, Dougal and Janet McIntyre, Robert McIntyre, David McConnell, Gordon Macadie, P R March, Mary Maxwell, Elsie Miller, Jimmy Mitchell, Bill Neill, Menzies Paterson, T D M Robertson, Robert Watt and Quentin Wilson.

Special thanks are due also to British Aerospace, Civil Aircraft Division at Prestwick for its assistance and above all to Jean Jackson, Erine Love, Iain Ghee and Iain MacLeod for their part in turning the manuscript into a finished product. Finally, the book would not have reached completion but for the constant support, advice and forbearance of my wife and family to whom it is gratefully dedicated.

ACR
1986

Preface

Half a century after its foundation is perhaps the ideal vantage point from which to review a company's history and particularly so in the present case, as there are many who can still recall at first hand the exciting new business of aviation to which they were attracted as young people in the 1930's. Now in 1986, which has been designated Industry Year, it is particularly appropriate to recollect the achievements of the intervening years and the contribution made by a comparatively small organisation not simply to the life of its local community but on the much wider stage of an international industry.

Researching this story has, however, been problematic since, inevitably in a business organisation, much of what will be tomorrow's history is today's waste-paper and is treated accordingly. But if documented archive materials are sparse, personal memory has proved to be a rich source and one of the objects of writing this history has been to take the opportunity, while it is still possible to do so, of obtaining and recording some of the personal insights and recollections of those who have been closely associated with the aircraft industry at Prestwick from its earliest days.

It may seem curious therefore that this account does not mention more people by name. This has been a conscious decision. Inevitably some individuals are identified, either because their role in the Company placed them centre-stage or because they happened to be involved in events which seemed to me at least to be particularly significant. Some have expressed the wish that they should not be singled out for mention from among so many others who deserve it. Rightly the name of David Fowler McIntyre occurs frequently, since from his part as co-founder in 1935 he was the driving force behind Scottish Aviation's development until his death in 1957 and, as an inspiration in the corporate memory, long beyond that. Generally, however, individuals remain un-named and if this means that I have done less than justice to a few, it also means that I have at least done justice to the anonymous many on whose endeavours the Company has ultimately depended in charting its course through fifty turbulent years.

There may be those who will be quick to object that it is wrong to speak of "the Company" in this way, since Scottish Aviation Limited itself dis not reach its half century and British Aerospace, into which it was merged, is not yet ten years old.

Scottish Aviation limited as such ceased to exist from 1st January 1978 but a change in legal identity is too artificial a break in the continuous fabric of the history of aviation at Prestwick. Of course there has been change in more than name alone since 1978, but the years before then are and always will be part of the story of what is now British Aerospace, Civil Aircraft Division - Prestwick. Equally, while the years since then belong first and foremost to the history of British Aerospace, they also belong to the history of the predecessor company, if only because they represent a consummation of so much of what Scottish Aviation set out to achieve.

LION RAMPANT and WINGED

G-ADWM
ADHN
ADYB

Chapter 1

"AB INITIO" Training

In 1933 two young pilots fired the imagination of the world when they became the first men to fly over the summit of Mount Everest. The two men were Douglas Douglas-Hamilton, Marquis of Clydesdale and David Fowler McIntyre. Using two specially converted Westland Wallace single-engined biplanes their breakthrough into sustained high altitude flying remains one of the pioneering feats of aviation.

It is not too fanciful perhaps to trace the conception of the Company which they founded two years later back to the success of the flight over Everest. Nowhere had their achievement been received with more enthusiasm and excitement than in their native Scotland and they openly acknowledged later that this receptive climate made a major contribution to the successful establishment of Scottish Aviation Limited.

Certainly, the objectives which they set for the new Company, "to establish the aviation industry in all its branches in Scotland," were scarcely less ambitious than those of the Everest Expedition itself. Heavily dependent on traditional industries the Scottish economy had suffered among the worst effects of the depression and needed the stimulus of new enterprises.

On the other hand, even in 1935 Scotland could already claim a historical involvement with aviation. Over four hundred years earlier King James IV had sponsored an enterprising abbot in his endeavour to fly by means of a pair of artificial wings and a gravity-assisted take-off from the rock of Stirling Castle. Unfortunately, the intrepid early aviator quickly found that gravity was more than a match for his invention and he was lucky to escape with a pair of broken legs after landing on a conveniently-placed dung heap.

More successful was the first balloon ascent from British soil which was made by the Scotsman James Tytler in Edinburgh in 1784 and, as David McIntyre used to like to point out, the first flight in Britain of a heavier-than-air machine had been made by a glider at Cardross in Dunbartonshire as early as 1895.

McIntyre himself came from a shipbuilding family, an industry in which the Scots had been particularly successful and aviation seemed to him to be not only a natural extension of Scotland's long involvement in shipping and shipbuilding but one which would have "a special appeal to the engineering and technical skill of the race."

Douglas, 14th Duke of Hamilton

David Fowler McIntyre

Lord Douglas Douglas-Hamilton, as Marquis of Clydesdale, was heir to Scotland's premier peerage, a family whose motto was "Never Behind" and whose members expected to set themselves demanding objectives. Among his ancestors was that Douglas who, charged with the task of carrying the heart of King Robert the Bruce to the Holy Land, had been ambushed by the Moors in Spain and, throwing the casket into the thick of the foe, had followed it to his death.

For Hamilton and McIntyre alike therefore it was natural for Scots to be in the fore. But it was one thing to imagine, in the heady altitudes of Mount Everest, the establishment of an aviation industry in all its branches in Scotland. It was quite another to turn this vision into reality.

Their first opportunity to do so came in the sphere of flying training. After the failure of the Geneva Disarmament Conference in 1934 it became clear that Britain would need to react swiftly to the growing menace of Nazi Germany's air force which had increased its published defence estimates for that year by 250%. In March the Prime Minister, Stanley Baldwin, responded to this ominous development by saying that "This Government will see to it that in air strength and air power this country shall no longer be in a position of any inferiority to any country within striking distance of our shores."

As a result, Parliament was informed that the strength of the Royal Air Force would be increased by forty one squadrons over the next five years to bring its total front-line strength to 128 squadrons, and before the end of the year this programme was further accelerated so that twenty-two of the squadrons reserved for home defence might be formed within two years.

In addition to a dramatic increase in the volume of aircraft production, this policy also required a major expansion in flying training and this requirement was not neglected, since the Air Staff consistently attached more importance to quality and performance than it did to mere numbers. In 1935, however, the Annual Budget was not geared to afford heavy capital expenditure on the creation of new training facilities. The Air Ministry therefore offered Royal Air Force training contracts under which private companies would put up their own capital for the construction of the airfields and buildings and for the purchase of aircraft and the other equipment necessary to provide training schools. In this way the Annual Budget had only to cover the current cost of the training contracts which included a small element to provide some return on the capital outlay of the private companies.

Scottish Aviation Limited in 1937

The political influence of the Hamilton family probably played some part in ensuring that a new Scottish company would receive serious consideration from the Air Ministry in tendering for a share of these contracts. At the same time it was Clydesdale's personal acquaintance with the Directors of the De Havilland Aircraft Company, as a result of his having ordered a Tiger Moth for his own use, which proved invaluable to the foundation of Scottish Aviation.

In offering training contracts, the Air Ministry had naturally approached established companies in the first instance and De Havilland itself set up schools at Maidenhead and Hatfield. Hamilton and McIntyre were therefore quickly able to interest D.H. in a collaborative venture in Scotland.

Consequently, when Number 12 Elementary Flying Training School was established at Prestwick, De Havilland provided not only the aircraft but also over a third of the initial capital making it the largest single shareholder. The original plan had been that the Hamilton Trustees would purchase the land on which the airfield was to be constructed and lease it to the Company but early in 1936 it was arranged that the Company itself should buy the property by means of a mortgage which made the Hamilton Estates the principal shareholder.

The De Havilland connection also provided Scottish Aviation with two of its original Directors. W.E. Nixon had been a founder member of D.H. in 1920 and was Financial Director when he became the Chairman of Scottish Aviation in 1935. He went on to become De Havillands' Managing Director and Chairman. T.P. Mills was Acting Chairman of D.H. when he joined the Scottish Aviation Board.

De Havilland Tiger Moth in Scottish Aviation livery.

The other Directors were R.L. Angus, a major local industrialist, David McIntyre and Lord Nigel Douglas-Hamilton, Douglas' younger brother, who represented the Hamilton family on the Board because Clydesdale himself was a Member of Parliament and as such was debarred from holding a Directorship in a company engaged on Government contracts.
The unique contribution which Clydesdale and McIntyre brought to the venture was ten years' familiarity with flying conditions in the West of Scotland, deriving from their experience as Commanding Officer and Deputy C.O. respectively of No. 602 (City of Glasgow) Squadron. Theirs was the recognition that the Prestwick area was characterised by quite exceptionally good weather and clear visibility conditions in a corridor some ten miles deep and twenty miles long.

To their astonishment, one of the first obstacles which they had to overcome was the Air Ministry's reluctance to accept Prestwick as the site for the proposed training school. Unfortunately one of the provisos set by the Ministry in granting the training contracts was the right to approve the location of the airfields. More unfortunately still, a report on its files stated that the Prestwick area was "exceptionally unfavourable for flying and flying training." The difficulty was eventually resolved when it was discovered that the record on file was the result of a comment by a former President of the Airfield Board, who, having been brought up in nearby Saltcoats, which is on the Greenock rain belt, had mistakenly assumed that the weather conditions prevalent there affected the West Coast in general.

Another early difficulty for the new Company was the selection of a name. Various alternatives were contemplated including "West of Scotland Aviation Limited," "Caledonian Aviation Limited" and "Clyde Aviation Limited". "Ayrshire Aviation Limited" and "North British Aviation Limited" were also considered but were not available. Nor, in the first instance, was the Company's preference, which was for "Scottish Aviation Limited," a name which would be indicative of the extent of its founders' aspirations. On investigating the availability of the various names from the Registrar of Joint Stock Companies it transpired that the name had previously belonged to another company which had since ceased to operate. In order to adopt it either the consent of that company's liquidator had to be obtained or else it had to be removed from the Register by statutory proceedings. The latter was obviously a difficult and protracted process but the consent of the liquidator was complicated by the fact that the individual concerned had died fifteen years earlier, since which time nothing further had been done in connection with the liquidation! Fortunately the firm to which he had belonged helped to arrange the appointment of a new Liquidator and this enabled the transaction to take place in March 1936.

In the meantime the name under which the Company was formally incorporated on 9th August 1935 was the "Scottish College of Aviation Limited", which was, as a notice board on the aerodrome fronting the main road went on to explain, a "Civil Training School for R.A.F. associated with the De Havilland Aircraft Company".

The rather elaborate lengths to which the notice board went to establish the Company's bona fides for the local public may at first seem curious, but it should be remembered that, modest as the original grass-strip and associated airfield facilities may seem by to-day's standards, in 1935 they represented a radical and not altogether welcome development.

The essentially rural nature of the community into which the aerodrome was thrust is well illustrated by one of the earliest decisions recorded in the Directors' Minute Book, when "the meeting confirmed the letting of the Grazings on the Company's Aerodrome to Mr. George Young, Aitkenbrae Farm, to Whitsunday 1936 at a rent of £10 per score of sheep".

157 acres of land were purchased to conform with the requirements of the current training standards, but again the Company made no secret of its longer-term plans and an additional 190 acres of adjoining land were simultaneously purchased with a view to the future extension of the airfield.

While David McIntyre and the Company Secretary, Stewart Kennedy, set up a first makeshift office in the bridal suite of the Orangefield Hotel, a gang of fencers arrived from the Hamilton and Kinneil Estates to mark out the ground and construction work on the airfield itself commenced in September 1935 with the removal of trees and hedgerows, the levelling of the ground and excavation for roadways, new drainage systems and the replacement of overhead with underground cables. This was followed by the construction of office and training accommodation together with a hangar and tarmac aprons to accommodate up to twenty aircraft.

Striking the first blows, 1935

Although the preparation work did not include the laying of concrete or tarmacadam runways, which were virtually unknown anywhere outside the United States before 1940, the establishment of the airfield nevertheless involved considerably more upheaval than simply the seeding and cultivation of a good grass surface. Even this process was dogged by the complications of poor weather in November which hindered rolling and eventually entailed the additional trouble and expense of a certain amount of turfing in order that the airfield might be ready to commence operations on schedule in February 1936.

Among those who were disturbed by the preparatory activities were the remains of some much earlier occupants of the site. In the course of excavation traces of burned bones together with Stone-Age burial urns, pottery, stone tools and flint knappers debris were unearthed in the remains of two cairns and an oval-hut site.

One contemporary view of the foundation of the airfield and its immediate impact in the locality can be traced in a poem entitled "Progress", which was written by a local teenager and published in the Evening News of March 7th 1936.

PROGRESS

"Here were fields, and here, the lovely trees,
A little time ago, when Spring's sweet air
Was breathed upon the country,
here and there
A bud unfurled to kiss the fragrant breeze -
Here were trees.

Hedgerows, that ran between the fields, were these,
A little time ago the brambles hung
Dew-black and ripe the thorny spikes among,
And in the air was murmuring of bees -
Here were trees.

Broken the hedge, and fallen are the trees,
A little time and none will ever know
That where these frames of steel and concrete grow
- A Martian crop upon the pleasant leas -
Here were trees.

Gone are the fields, star-sprinkled in the breeze,
A little time and who will call to mind
That where these humming giants churn and grind
And swoop as hawks their simple prey to seize,
Here were trees?"

Elsie Wightman 1936

"These humming giants" Blenheim, Anson and Tiger Moths.

Others were less subtle in their criticism and at one stage in the construction opposition reached such a pitch that work had to be stopped and the contractors escorted from the site by the police. In these circumstances the Company had to draw attention to the future advantages which it believed would accrue to the neighbourhood from the development of the airfield and from the ancillary industries and air traffic which it could be expected to generate. Acceptance of the project was eventually achieved only at a special meeting of the Ayr Town Planning Committee.

Nevertheless by the middle of March 1936, when flying training had been in operation for no more than a month, the Company and its Directors had received various letters and personal complaints objecting to the noise of its aircraft. By the time the trainees of the third course were flying their first solos in July the idea of fitting longer and perforated exhausts was being explored as a means of countering this problem.

In these circumstances it was hardly surprising that the Company remained sensitive to local opinion. It was particularly conscious of the difficulties associated with flying on Sundays. In May 1936 the Company received an unofficial report that the Air Ministry might seek Sunday Flying in order to expand the available training capacity. Aware of the particular opposition which this might encounter in Scotland it was decided that should there be any official intimation of this requirement, then a Board Meeting should immediately be convened, "as such a step would require to be carefully considered before any action could be taken".

In August 1936 the Company was indeed formally invited to tender for week-end training of the R.A.F. Volunteer Reserve. While it was anxious that declining to tender might be prejudicial to the Company's ability to secure further contracts, the Board remained apprehensive about Sunday flying. The alternative of simply ceasing to fly during Church Service hours was discussed, but eventually the Company pointed out to the Air Ministry that, in view of the particular importance attached to religious observation in Scotland, it intended to complete the training without doing any flying on Sunday mornings. Even with Prestwick's good weather conditions this made the weekend training programme very tight and the pilots used to listen for the sound of the church bells announcing the end of the morning service to give them the signal to start their engines.

Expansion - Tiger Moths in front of the "Tiger" Hangar with the extended administration block and the new "Anson" Hangar in the background, 1939

The Company's first contract with the Air Ministry had been signed on 14th October 1935. Under this contract the Company undertook to provide the first 50 hours of 'ab initio' instruction to civilian trainees prior to their joining the R.A.F. The contract also stipulated that No. 12 E.F.T.S. at Prestwick was to be ready to receive pupils on 17th February 1936, bringing the date set out in the original tender forward by two weeks.

The urgency of the Government's efforts to expand flying training had already been firmly impressed on Clydesdale and McIntyre through their experience in submitting the Scottish Aviation tender. When they had received the tender document itself from the Ministry in London they found that the deadline for their submission was such that they did not even have time to return to Prestwick to complete it. Discussing this problem as they travelled across London in a taxi, they suddenly spotted a commercial delivery van and instructed their driver to follow it in the hope that this would lead them to an office where they could try to borrow a typewriter. By all accounts both were men of great personal charm. They duly succeeded in obtaining this favour from a total stranger and the Scottish Aviation tender was typed and returned to the Ministry on time.

The first contract not only brought forward the starting date of the training, it also increased the scheduled programme to more than double the Company's original expectations. At a level of over 8000 hours it became necessary to acquire a total of twenty Tiger Moths from

De Havilland rather than the sixteen which had been originally planned. The aircraft themselves were painted in a striking livery, the fuselages a flame orange - a favourite colour of McIntyre's which was to be employed repeatedly under his direction - with ivory-coloured lettering and silver wings. In addition to supplying the aircraft, the De Havilland connection gave the Company the opportunity to acquire its petrol through a Special Bonus Agreement at a price of one shilling and one sixteenth of a penny per gallon. The Scottish Aviation oil contract was awarded to Duckhams at a price of one shilling and ten pence per gallon.

The Flying Training Instructors

The Ground Engineers

By the end of November seven of the establishment of eight Flying Instructors had been engaged and most of the important Ground and Maintenance Staff positions had been filled except that of Chief Ground Engineer. This post was eventually filled by F.W. Hopkins who came to Scottish Aviation from De Havilland where he had been Deputy Service Manager. Whenever possible it was men like this, whose abilities were already known, who were selected to form the nucleus of the new team. Some had taken part in the Everest Expedition, like R.C.W. Ellison, who had been the R.A.F. Liaison Officer responsible for cloud investigation and F.W. Burnard, the Westland Engineer. Others had served with 602 Squadron or, like N.J. Capper, knew McIntyre from his period with No. 12 Tactical Bombing Squadron. While it was inevitably necessary to recruit personnel from outside the area for these specialist positions, a conscious policy was adopted to fill all non-technical posts with local labour.

Only one of the Flying Instructors had any previous experience of giving instruction, but all were experienced pilots. Their pay was made up of a retaining fee of £300 per annum with a guaranteed minimum of £500 made up through flying pay at a rate of ten shillings per hour. The Chief Instructor enjoyed a retainer of £400 and a guaranteed minimum of £650 made up at the same rate. The question of holidays was a difficult one in view of the arduous programme of training which had to be tackled but it was left to the Managing Director to ensure that the Instructors were given as much leave as possible. McIntyre himself took his share of pupils and acted briefly as Chief Flying Training Instructor until H.A. Barrow was appointed.

Keeping them flying, 1938 . . . . . and, for J. McGowan (centre), still doing so in 1985.

The first course commenced in mid February 1936 with a total of thirty-four pupils. It immediately became apparent that the founders' confidence in the weather conditions at Prestwick were entirely justified and that training could take place virtually without interruption and with much greater regularity than at the schools located in other parts of Britain. The log-book of one of the original Instructors reveals that in the whole of 1936 only three days were lost to flying as a result of adverse weather conditions and these were on the 3rd, 4th and 16th of December, as a result of very strong gales.

The first course was completed early, in little more than half the allotted time, and raised the Company's hopes that it might as a result obtain further hours or an additional allocation of pupils. In the event Scottish Aviation completed nearly 9,000 hours in 1936 by comparison with an original expectation of 3,200 and the Instructors were still able to enjoy a very generous allocation of some eleven weeks' leave during the year. Prestwick's advantageous position was quickly recognised and when in December 1936, the Air Ministry announced the total number of training hours which it planned to allocate to the various Schools under a new three-year contract, the figures awarded to No. 12 E.F.T.S. at Prestwick were the highest of all at 22,150 "ab initio" hours and 7,380 hours on Service types. As it transpired, Prestwick actually completed nearly 50,000 hours of pilot training during the three-year period.

Trainees were drawn originally from the Auxiliary Air Force and the R.A.F. Volunteer Reserve. The Auxiliary Air Force had been created as early as 1925 as a result of legislation passed in the previous year. It originally consisted of four squadrons but by the outbreak of war in September 1939 it comprised 20 squadrons of aircraft and 44 of balloons. Its officers and men were recruited for a minimum period of five years and drawn from the locality by the Territorial Association, in precisely the same way as the Territorial Army.

The Royal Air Force Volunteer Reserve was created in July 1936. It too was composed of civilians recruited in the neighbourhood of aerodromes and confined to men between the ages of eighteen and twenty five. Originally designed only for air crews, it became, when war broke out, the principal means of entry into the R.A.F..

Spotting their target? . . . . Hart and Audax aircraft from No. 12 E.F.T.S. flying over the Palace of Engineering in its original location in Bellahouston Park, Glasgow during the Empire Exhibition of 1938. Two years later it was brought to Prestwick to become No. 3 Factory!

Competition to become a pilot in the Auxiliary Air Force and the Volunteer Reserve was intense and the publicity surrounding the Scottish Conquerors of Everest, together with a series of Empire Air Day displays, ensured that No. 12 E.F.T.S. had a particular lure. So far as the young men who aspired to train at Prestwick were concerned, the qualifications which they required even to stand a chance of an interview to become a Sergeant Pilot in the Volunteer Reserve were a 'group' pass in Scottish Highers, that is, Higher English and Maths and two Lowers. They were paid one shilling per hour attendance money and other than a two-week residential spell in the summer were typically at Prestwick for training from 1 o'clock until 5 pm on Saturdays and Sundays. They also received their bus fares from Glasgow, although, as many were quick to discover, it was more economic for them to squeeze five together into a Morris 8 and claim 5 bus fares. Most of the trainees stayed overnight finding accommodation in the lofts of private bungalows on the northern edge of Prestwick, a few minutes' walk from the airfield. In 1937 these provided a Saturday evening meal, bed and Sunday breakfast for seven shillings and sixpence.

The sequence of instruction was carefully laid down, as follows:-

1. Air experience for pupils.
2. Effect of controls
3. Taxying and handling of engine.
4. Straight and level flying.
5. Stalling, climbing and gliding.
6. Medium turns, with and without engine.
7. Taking off into wind.
8. Approaches and landings.
9. Preliminary action in the event of fire.
10. Spinning.
11. First solo.
12. Side slipping
13. Further action in the event of fire.
14. Low flying (with instruction only).
15. Steep turns, with and without engine.
16. Climbing turns.
17. Forced landings.
18. Landing with engine.
19. Instrument flying under hood.

20a. Stopping and starting of engine in the air.

20. Taking off and landing across wind.
21. Aerobatics.
22. Air navigation.
23. Forced landing test.
24. Height test
25. Cross country test.
26. Passenger test.
27. Night flying.

Lesson 17?

The programme of training was intensive and the regime for the civilian trainees was strict. McIntyre whose hair had already turned its distinctive silver-grey, was an imposing and slightly awesome figure. Trainees commented that they found themselves "virtually under open arrest" at Prestwick but discovered that the discipline stood them in good stead when they went on to join the R.A.F.

At the same time instructors and pupils alike were characterised by a tremendous enthusiasm for what they were doing. Their recollections are of a unique combination of hard work and laughter, a sense of adventure and excitement.

The outcome spoke for itself and both the Central Flying School Report and the Superintendent of Reserves saw fit to comment in writing on the highly satisfactory results of Prestwick's training.

Empire Air Day at Prestwick, 1937

Through the flying log of one of the early Volunteer Reserve trainees it is possible to gain an insight into the progress of a young pilot at No. 12 E.F.T.S.. Sergeant Pilot (subsequently Wing Commander) Menzies Paterson embarked on the first stages of his training, "air experience for pupils" and "effect of controls" in a forty minute flight in Tiger Moth L.6949 on October 23rd, 1937. Travelling from Hamilton each weekend he had a total of 12 hours and twenty minutes experience when he made his first fifteen-minute solo in the same aircraft on December 18th, in the excitement of which, in his own words, he "forgot to pull back the 'cheese-cutter' and had to go round again," landing the second time with two inches of snow on the grass.

By the second half of February 1938 he had accumulated a total of 32 hours 40 minutes dual and solo, had progressed to aerobatics and was ready for R.A.F. Test No. 1, an occasion which was not made any easier by the fact that his examiner, Flight Lieutenant Ling, seated in the front cockpit was 6'3" tall and difficult to see past!

Cross country flights, when they came, were commonly made over three sectors encompassing Prestwick, Abbotsinch and Dungavel, the Duke of Hamilton's estate in Lanarkshire.

Going solo

In the middle of June 1938, with a total of just under 68 hours on the "elementary" Tiger Moths, he went on to the Service types which had by now been added to the training fleet at Prestwick, flying Hawker Hart K-6547 with Flight Lieutenant Capper. By the end of July, by which time he had his wings, he had progressed from Harts and Hinds on to "the big stuff" in the form of the Fairey Battle. On this machine he found the roles, and associated problems, which he had experienced on R.A.F. Test No. 1, were now reversed and his first solo came after only one hour and ten minutes flying 'dual' because the poor instructor in the back cockpit could not see what was going on anyway!.

On Friday September 1st, 1939, he finished his Volunteer Reserve training at Prestwick practising low flying and aerobatics in a Hind. The next day he was called up and, as one of the first Prestwick Volunteer Reserve pilots to be transferred under an agreement made earlier between the Air Ministry and the Royal Navy, found himself with three others in the Fleet Air Arm at Donibristle on the Firth of Forth on the Sunday when war was declared. He had completed a total of 62 hours dual and 135 hours 45 minutes solo flying during his 22 months at No. 12 E.F.T.S. and he went on to pursue a distinguished flying career.

The expanded school and office facilities viewed from the Sandyford Road, c.1939

Continuing expansion. The aerodrome from the Prestwick side c. 1939 showing the new "Fokker" hangar on the right.

When he subsequently mistook a whale for an enemy submarine he could perhaps be excused on the grounds that this particular line of identification had been one of the few areas of flying training which was not sooner or later undertaken at Prestwick. What was remarkable about the Scottish Aviation School was not simply the absolute volume of flying instruction which it eventually completed but also the range of this training.

The initial work of the School had included the provision of Flying Instructors' and instrument F .ying courses. Weekend training of the R.A.F. Volunteer Reserve was added from April 1937. In May the following year the Company erected a recruiting notice for the R.A.F.V.R. on the main Ayr road and this contract eventually built up to 260 pupils.

In February 1938 the Air Ministry approached Scottish Aviation with two new proposals. The first was to supply flying training for R.A.F. regulars and the second was to set up an Air Observer Navigation School. The Company quickly expressed its willingness to assist. Within three months consideration was being given to doubling the number of courses for regular "ab initio" pupils and in September it was agreed that S.A.L. should provide a further 588 hours per annum of this training.

Meanwhile No. 1 A.O.N.S. had commenced operations at Prestwick on 15th August with an initial complement of thirty. Additional courses soon followed and a year later the School was providing for 390 pupils under the direction of J.A. Dobson.

Finally, with the outbreak of war in September 1939, Wireless Operator training was added to Scottish Aviation's repertoire with the foundation of No. 2 Supplementary Wireless School, actually dispersed over three large local houses, Adamton, Rosemount and Dalkeith which the Company had requisitioned. Accommodation was not easy for 360 pupils, a total which very rapidly increased to over 400.

Space on the ground was not the only problem. The very rapid expansion which had taken place in the Company's training activities had made it necessary for No. 12 E.F.T.S. to increase the number of its Tiger Moths to 45 by July 1940 and obviously there had been a corresponding growth in the number of Instructors. That same month the Company had to increase the teaching establishment at No. 1 A.O.N.S. by a further four Navigational Instructors, one Photographic Instructor and four Assistants.

One of the ways in which the Company sought to alleviate the shortage of Instructors was through the novel development of "flying classrooms" for the simultaneous training of several pupils.

The aircraft used for this purpose were Avro Ansons and three 4-engined Fokker airliners which had been acquired from K.L.M. and flown out of Holland literally on the eve of its occupation by the Germans. One of the aircraft was an F.XXXVI and the others were F.XXII Types. The former, G-AFZR, was eventually burnt out after a take-off accident at Prestwick in May 1941, while G-AFXR, inelegantly named "Brontosaurus", finally crashed near Loch Tarbert in July 1943 after an in-flight fire, but not before becoming involved in a celebrated piece of Scottish folklore when it carried some of the supplies from the shipwreck which subsequently inspired Compton Mackenzie's book "Whisky Galore". The third aircraft, G-AFZP, was restored to a civil airliner role by Scottish Aviation after the war and was ultimately scrapped in 1952.

"The flying classroom"......... Fokker airliner in uniform

Like most other innovations the flying classroom method had its critics. A visit to Prestwick by the Air Officer Commanding Reserve Command early in 1940 had suggested that the authorities took an unfavourable view of this approach to training and were likely to take steps to withdraw the Air Ministry's approval of it. David McIntyre, however, sprang to its defence with his customary vigour, setting out the case for the "Fokker-Anson" method in a letter to the Under-Secretary of State for Air and asking for an immediate investigation to be made so that full consideration might be given to introducing this method of Air Observer training more widely.

By this stage the curtain was coming down on training at Prestwick, and consequently on the first chapter in the history of Scottish Aviation, following the decision to transfer these activities progressively from Great Britain itself to Canada and South Africa under the Empire Training Scheme.

No. 2 S.W.S. closed in Mid-November 1940, and No. 12 E.F.T.S. on 22nd March 1941, although not before some consideration had been given to developing the airfield at Turnberry to allow it to continue in Prestwick's good weather corridor. Air Observer training survived flying training at Prestwick by a few months and did not stop until 19th July, 1941.

Nor should S.A.L.'s other training locations be forgotten. The Air Observer School (No. 4 A.O.N.S.) which the Company had run under a Management contract at Dumfries from 4th October 1941 and which had at one time employed 284 staff, eventually closed in June 1942. The airfield which Scottish Aviation opened at Grangemouth on 1st May 1939 accommodated No. 35 Elementary and Reserve Flying Training School and No. 10 Civil Air Navigation School. Both were administered by the Company, although the former was closed soon after the outbreak of the war and the latter was merged into No. 1 A.O.N.S. at Prestwick in December of the same year as a result of enemy air raids.

Training had been a profitable activity for the Company and not only in a financial sense, although in May 1936 the Directors had been faced with a problem which perhaps occupied them all too seldom in later years, namely how to invest the Company's surplus of liquid cash. It had also been profitable in terms of the experience and above all the reputation acquired during this period.

Scottish Aviation's record as a training organisation had proved second to none. In marked contrast to flying training during the First World War, the only fatal accident during training at SAL occurred on 10 January 1939 when an Anson from the Air Observer School went missing and crashed into a hillside on the Rhinns of Kells in Kirkcudbrightshire killing the crew of four. Among the aircraft scrambled to look for the missing Anson the next day were the Tiger Moths of No 12 EFTS and by a strange coincidence one of these was caught by a phenomenon with which Hamilton and McIntyre had become all too familiar in the Himalayan mountains, a sudden down-draught of air. The Tiger Moth crashed within 300 yards of the Anson but its pilot and observer were more fortunate and escaped shaken but unhurt.

During the five years of its operational existence between February 1936 and March 1941 No. 12 Elementary Flying Training School had passed out 1,334 pilots and completed a total of 88,477 hours and 25 minutes of flying training. No. 1 Air Observer Navigation School had flown an aggregate of 134,530 hours and had produced in the process 1,994 Air Observers, over half the total in the Royal Air Force at that time. In its brief year No. 2 Supplementary Wireless School had trained nearly 1,200 Wireless Operators. By the time that flying training was moved as a matter of Government policy to the Empire Air Training Scheme in 1941 Scottish Aviation was training five times as many Royal Air Force personnel as any other company undertaking R.A.F. training contracts in Great Britain.

Despite the scale of this achievement training had already become a relatively small part of the Company's activities. These had moved progressively from the men who flew to the aircraft themselves and to the establishment of another branch of the aviation industry in Scotland.

"The discipline stood them in good stead"...... trainees of No. 12 E.F.T.S. flying in close formation.

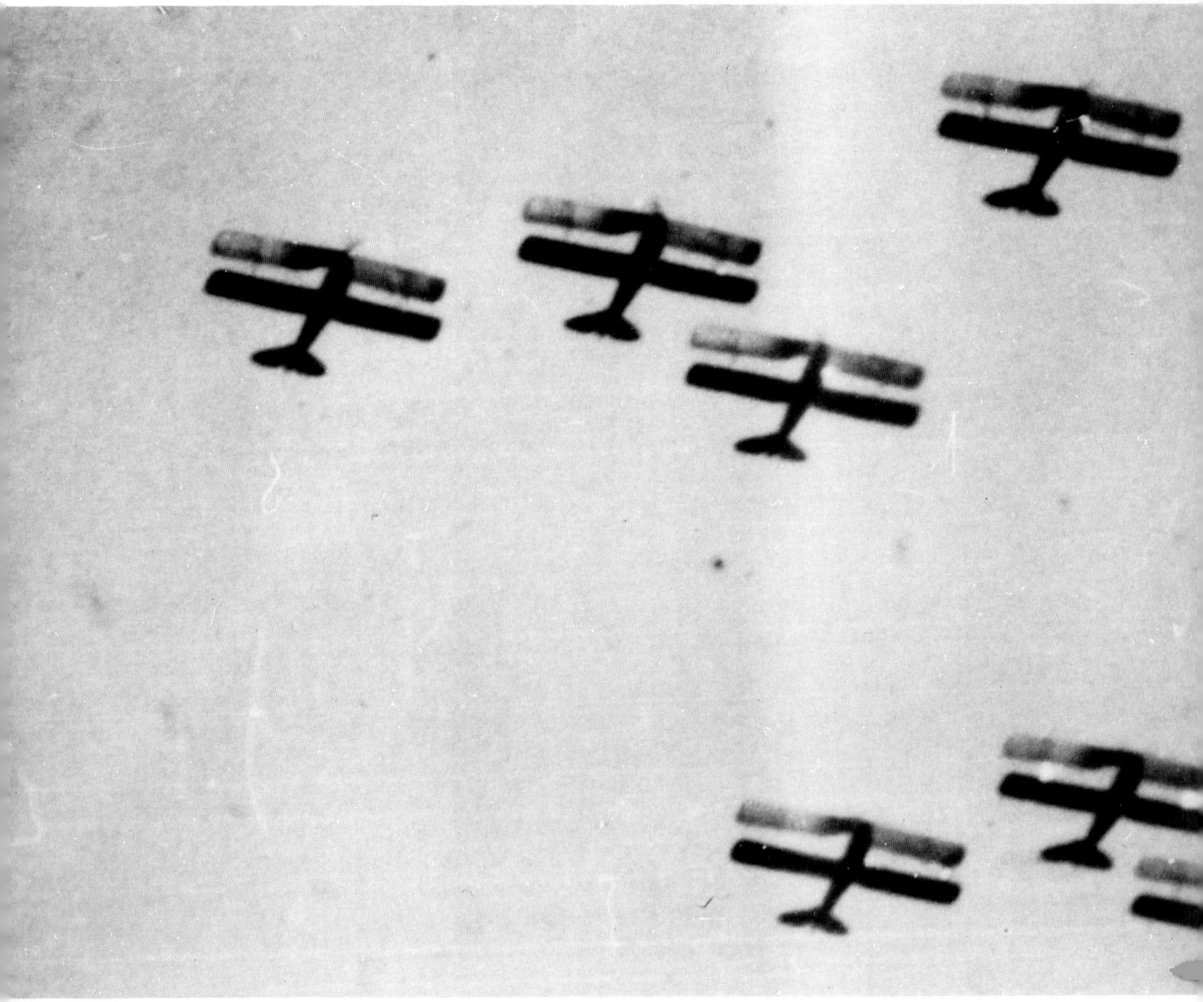

Chapter 2

"THE STIMULUS of WAR" Aircraft Engineering

The ability to overhaul and maintain its own aircraft was of course an integral feature of the Company's flying training organisation and, in view of the founders' objectives, it was not long before Scottish Aviation was seeking to set up its own aircraft manufacturing capability.

In pursuit of this aim approaches were made to other British manufacturers and to the Government departments responsible for aircraft supply in an attempt to obtain sub-contract orders for the manufacture of parts.

At first it seemed as if the intimate connection with the De Havilland Aircraft Company which had been instrumental in the establishment of the flying training school would also give Scottish Aviation its initial breakthrough into aircraft production. De Havilland itself was obtaining substantial new contracts from the Air Ministry and as early as July 1936 was exploring with SAL the possibility of setting up a workshop adjacent to Newdykes Farm to undertake the manufacture of metal propellers under sub-contract. At the same time, however, De Havilland was also considering the alternative of establishing a new works of its own in Lancashire, where Bolton Corporation was offering sites on an extremely attractive basis in an effort to reduce high local unemployment. As a result the propeller contract eventually went to Bolton and gave birth to DH Lostock.

It might have been expected that, as the pace of rearmament accelerated, Government contracts would prove relatively easy to secure. However, as in its earlier efforts to persuade officialdom of Prestwick's suitability as a training location, the Company now encountered unexpected obstacles in obtaining manufacturing work.

To some extent perhaps, this was an inevitable difficulty for a new firm with an unproven record in this sphere, but it seems that there were other inhibiting factors too. In

September 1938 Lord Nigel Douglas-Hamilton, Earl of Selkirk, received a letter from the Air Ministry stating that it considered the Prestwick site unsuitable for the purpose of establishing an aircraft factory.

The reasons for this view are not clear but it appears that the Company may have been handicapped by an expectation within certain Government departments in London that the troubled industrial relations of Glasgow's Clydeside would also be characteristic of the Prestwick area. At one meeting which Hamilton and McIntyre attended, they were told by a senior official that he personally would on no account approve any work connected with the supply of aircraft or other equipment for the fighting services being allotted to "Red Clydeside".

Despite the early disappointments the Scottish Aviation Board was unanimous that it was not prepared to achieve its entry into manufacturing at the expense of an external association which might prove to be prejudicial to its own independence. For this reason it was agreed in September 1938 that earlier discussions regarding an association with Rollason Aircraft Services Limited of Croydon should not be pursued.

SAL was determined to persevere in its endeavour to establish a manufacturing facility and, reinvesting the profits made from its training activities, went ahead on a speculative basis to construct and equip its first factory for repair and modification work on a site behind Orangefield Mains Farm.

The gamble paid off. Shortly afterwards, early in 1939, the Company secured a series of orders for its new factory. Ironically the first was modification work on Wellesley aircraft which came to Scottish as a result of an overload at Rollasons. By June of that year the Air Ministry and A.I.D. were expressing their satisfaction with the way in which the work had been carried out, the time taken comparing very favourably with that of other sub-contractors. As a result, the Company was informed that it would be allotted 35 aircraft to modify, acting as a "fringe" firm to the Blackburn Aircraft Company based at Brough. This turned out to be work on the Skua dive-bomber and its variant, the Roc. The first of these aircraft arrived at Prestwick on 2nd October 1939 and the last was despatched on 22nd April 1941.

In terms of new manufacture Scottish Aviation's first undertaking was the production of Hart and Hurricane rudders for the Hawker Aircraft Company. Initially the Company's inexperience was all too evident. Surveying

Prestwick's woefully inadequate provision of tools and fixtures for the new contract prompted one Hawker man to declare that they were not sufficient to make a wheelbarrow, never mind an aircraft. In May 1939 the Superintendent who had been appointed to take charge of the programme was therefore despatched to Kingston to familiarise himself with the task and Scottish applied itself both diligently and cannily. The second rudder to arrive back at Kingston purporting to be an SAL product turned out still to have its Hawker Aircraft inspection stamps on it, being a specimen which had originally been sent North for instructional purposes. In any event Scottish learned quickly and in the first twelve months of production 350 Hurricane rudders were made at Prestwick.

The principal problem with the early contracts was the difficulty of securing continuity of work for the new facility. A lengthy idle period followed the completion of the Wellesley modifications before the commencement of the Blackburn work and it was the outbreak of war which finally provided the stimulus that enabled the Company to establish its manufacturing interests on an ongoing basis and a significant scale.

Within a month of the declaration of hostilities the Company was asked to undertake production of the Lysander, the light communications aircraft designed by Westland and subsequently most famous for its "moonlighter" role, ferrying agents to and from occupied Europe. The initial proposal was that production should rise to a rate of eight aircraft per week by the end of 1940. It was recognised that the scale of this task was such that 55% of the work would require to be sub-contracted from Prestwick, but it was to be left to Scottish Aviation to arrange the placing of all the sub-contracts. The Air Ministry would put up all the capital required for building, plant and tools although the Company was required to provide working capital.

Westland Lysander at Prestwick, 1939

This was clearly a major opportunity for Scottish Aviation to establish itself as a manufacturer in its own right, but it was not without its problems. One estimate of the working capital required for the proposed Lysander programme was as high as £250,000, a figure several times greater than the Company's original capitalisation. In the circumstances a certain tension between the ambitions of the Scottish founders and the interests of their De Havilland colleagues now came to the surface and at a Board Meeting in London in November 1939 Mr. W.E. Nixon revealed that the De Havilland Board had decided that it must ask to be relieved of its interest in SAL.

He explained that the geographic separation between the two companies and the heavy workload on De Havilland itself at that time were felt to make it impossible for it to be able to provide SAL with the support, both in terms of expertise and - implicitly - finance, which would be necessary to embark on such a large-scale manufacturing programme.

Various alternatives were considered in an attempt to resolve this problem, including the proposal that the Air Ministry should be approached with a view to supplying the working capital under a "Shadow Factory" scheme and that Scottish Aviation should raise at least part of the necessary finance by realising other assets through the sale of the aerodrome which it had established at Grangemouth. The latter had only been open since the summer of 1939 and it was still at this time the aspiration of David McIntyre and the Hamiltons that it might, in due course, become the civil airport to serve the whole of Central Scotland and it was natural therefore that they were reluctant to pursue this second alternative.

The matter was eventually resolved to the satisfaction of the De Havilland Directors in January 1940 when it was agreed that SAL would undertake the Lysander work not as prime contractor to the Ministry but as a sub-contractor to Westland. Even so, the demands on the Company's resources were considerable. It was decided that the balance of unissued capital should be called up at the end of February and the Company's overdraft facilities were extended from £120,000 to £170,000. The stimulus of war was proving to be a double-edged sword.

The Lysander project suddenly presented the Company with the prospect of an acute shortage of factory space. David McIntyre had, however, already identified an opportunity to resolve this situation quickly. His dramatic solution was to purchase a second-hand building and have it transferred to Prestwick.

The building in question was The Palace of Engineering which stood at the time in Bellahouston Park in Glasgow where it had been constructed to accommodate the engineering exhibitors at the Empire Exhibition of 1938. During that time it had housed the stands of many famous companies including John Brown's, G.E.C., and Rolls Royce, whose engines were to be worked upon by Scottish Aviation for many years in the same building after its move to its new home at Prestwick.

Site of "The Palace" at Prestwick 25.4.40

3.6.40

15.6.40

6.7.40

30.7.40

12.8.40

2.9.40

16.9.40

In January 1940, the Air Ministry approved the Company's plan and its proposed site and in only four months the Palace had been dismantled, transported to Prestwick and re-erected. Despite the fact that it now represented by far the largest single workshop area and, particularly after the disastrous fire of February 1941 which destroyed the original administration block, provided the bulk of the Company's office accommodation, its official designation at this stage was "No. 3 Factory", which put it firmly in its place behind its smaller, but senior, counterparts, "The Tiger" and "The Anson" Hangars. Although it was only five years old, the Company had already established certain traditions.

The speed with which this additional factory facility was made operational was due in no small part to the efforts of the Company's own Civil Engineering Department which had been formed to manage the problem of accommodation which had accompanied SAL's rapid growth. Outside contractors were also utilised and, as in the case of their earlier involvement with the original construction of the airfield, their presence was not without incident. Among the contractors who were employed in raising the Palace was a team of Canadian ice-hockey players, who had the habit of working in their numbered jerseys. Whether this was intended to foster team spirit is not clear but in view of their common appearance the site Foreman used to distinguish them by their numbers rather than their faces.

Deciding eventually that the performance of certain men was not satisfactory, he called out their numbers and instructed the Chargehand to dismiss them from the site. With all labour in scarce supply because of the war effort, however, the Company was reluctant to see its programme slipping as a result of the men's departure and by the ingenious removal of the offending jerseys and some judicious re-numbering the gang was kept up to strength.

In the same way as Scottish Aviation's training activities had come to extend beyond Prestwick, so it was with its manufacturing interests. In March 1940 the Company had received a tentative approach from the Air Ministry regarding the possibility of its undertaking the repair of flying boats at a Clydeside shipyard, since it was believed that this might be of some interest as a joint venture for the shipbuilding companies. It had by this time been recognised that the whole of Britain's flying boat capacity lay on the South and South-East coasts of England and would therefore be within easy range of enemy bombers in the event of France and Belgium being invaded.

The flying boat manufacturers Saunders Roe, for instance, were looking for an alternative site to which they could transfer their Cowes works, and, after one of their Directors had visited Prestwick, Scottish Aviation put forward the suggestion of an association between the two companies. Consideration was given to the formation of a new company to operate flying boat repairs at Greenock and, with a view to strengthening the Scottish interests in the project, an approach was made to shipbuilder Sir James Lithgow to see if he was prepared to participate, although in the event other commitments prevented him from being able to do so.

Sal, representing Scottish Aviation's women at war, as she appeared in the Greenock factory's newsletter.

Short Sunderland flying boat under repair at Greenock.

In any event, Scottish Aviation's preference was to operate Greenock itself and to undertake sub-contract work for Saunders-Roe rather than share the site. The particular yard in Greenock which the Company was seeking to acquire was Caird's, established in the nineteenth century on the site of the graveyard where Robert Burns' Highland Mary was buried. Negotiations for its acquisition were protracted and difficult owing to the fact that the Admiralty was also taking an interest in the site for the expansion of a torpedo plant. In the end Scottish Aviation persuaded the Air Ministry to forestall the Admiralty by requisitioning the site, which was particularly well located for the operation of flying boats. Lord Beaverbrook himself, then Minister of Aircaft Production, came to visit the site, agreed with the Company's view as to its suitability and instructed that work on its conversion should commence immediately. Impressed by a decisiveness which he found all too often lacking in his dealings with politicians over the years, David McIntyre inquired how the Minister managed to persuade the Treasury to sanction the spending so quickly. The explanation, as Beaverbrook revealed, was not to let the Treasury know until the work was well in hand. Caird's was only one of half a dozen sites which he then had in preparation.
McIntyre had no cause to complain of political prevarication on this occasion. The yard was formally requisitioned on 1st June 1940 and ten days later Scottish Aviation received a contract to oversee its transformation into a flying boat base and to be responsible for the repair and overhaul at that site of various British types, notably Sunderlands.

As at Prestwick, the workload at Greenock was subject to short-term fluctuation and sudden contingencies throughout the War, but the overall load was such that a further Scottish Aviation outstation eventually had to be established at Largs. This opened originally as a salvage section and machine shop in February 1942 but within a month it was decided that work should proceed to add a slipway and, when this was completed in September of that year, Largs' primary role became the provision of dispersal capacity for the base at Greenock.

By far the most unusual of Scottish Aviation's satellite manufacturing operations during the War was that located in an old warehouse at 39 West Campbell Street in Glasgow.

Curiously it was here rather than at Prestwick that the Company undertook the production of its first complete aircraft, the "Queen Bee", a pilotless radio-controlled variant of the DH82 Tiger Moth which was used for aerial gunnery practice.

The preliminary plans made for this project in September 1942 envisaged production of 150 aircraft, the first of which was being cleared by De Havilland one year later. In December 1943, however, the Ministry of Aircraft Production cut the programme back to 60 units and simultaneously deferred the delivery dates.

Flight testing of the aircraft was carried out at Prestwick and the product was either robustly constructed or the gunners were poor marksmen, because 43 of the aircraft which had been produced by SAL were offered on the civil market after the War as a source of Tiger Moth spares. Perhaps the most interesting feature of the project sprang from the fact that the aircraft was essentially a wooden construction and as such Scottish Aviation selected a furniture manufacturer, Messrs Morris and the West of Scotland Furniture Company from Beith to act as its principal sub-contractor.

At Prestwick itself the engineering programmes had gone through a major upheaval during the summer of 1940 as aircraft production capacity throughout the country was applied to the task of strengthening the fighter-squadrons which were shortly to be called upon to defend Britain against German invasion. Although further orders for Hurricane rudders were received, repair and modification of the Blackburn Rocs and Skuas were reallocated to the Admiralty and most significantly the proposed Lysander project was abruptly cancelled except for a limited amount of component manufacture. Instead, under the direction of the Civilian Repair Organisation based at Merton College, Oxford, the Company became a primary site for the repair of fighters, some Hurricanes but predominantly Spitfires, the plan being to develop throughput to a rate of twelve aircraft per week. Consideration was also given to the production of certain components as a sub-contract to Supermarine with a view to Prestwick eventually being in a position where it could become capable of undertaking the complete manufacture of Spitfires at short notice.

SAL also acted as parent to another local civilian organisation which had been given the job of Spitfire repairs, namely the London, Midland and Scottish Railway's wagon works at Barassie. Aircraft repaired there underwent test flight and final inspection at Prestwick.

The volume of work involved on Spitfire repairs was such that it more than made up for the loss of the Lysander production and, although the programme fluctuated, it remained a major part of the Company's engineering activities until June 1945 with a normal holding of sixty to seventy aircraft in the factory. During this time some 1,200 Spitfires were repaired and returned to service from Prestwick.

Those girls again! On Spitfires at Prestwick

It was in connection with Spitfire repairs that one of the most bizarre episodes of the war years took place. Having been overhauled, the aircraft were taken for engine ground running prior to test flight. During the ground running process two engineers would lean over the tail to hold it down, facing backwards to avoid the slipstream. On a pre-determined signal they would release the tail and step away when the pilot was ready to take off. On this occasion, however, one of the two failed to notice the signal and, before he could do anything about it, found himself airborne, still clinging to the tail. The pilot, finding the controls sluggish and unresponsive, glanced in his mirror only to encounter the shock of discovering his unexpected passenger. He carefully turned and brought the aircraft round into a position from which it could land and although unfortunately the ground engineer was unable to maintain his hold during the final approach and was badly injured in his fall, he was lucky enough to live to tell the tale.

Until the middle of 1940, Scottish Aviation's involvement in aircraft engineering had been the traditional one of a sub-contractor to other British aircraft manufacturers and their types. Then, suddenly, its role was to change in a way which was to have far-reaching implications for the development of the Company and of Prestwick Airport both during and after the War.

The change came with the advent of the Lease-Lend Agreement under which American aircraft and other equipment were supplied to Britain even before the United States had entered the War. The Company's initial involvement with this process came in July 1940 when it was asked to undertake the assembly at Abbotsinch of sixteen aircraft which had arrived from America.

This was followed by the assembly of further Grumman, Northrop, Chance Vought and Curtiss fighters including Tomahawks, Mohawks, Kittyhawks, Kingfishers and Martlets, but the event which was ultimately to have the most profound significance came on 29th November when a Lockheed Hudson aircraft landed at Prestwick having crossed the Atlantic from Newfoundland. The pilot had in fact been scheduled to land at Aldergrove in Northern Ireland which had served as the U.K. terminal for these trans-Atlantic flights since they had begun, although that was less than three weeks earlier, but he had become separated from the remainder of his formation and had landed instead in Scotland. Thereafter and for the remainder of the War, Prestwick became the principal terminal for all land-planes flying from North America and it became Scottish Aviation's responsibility to service and modify these American types to British operational standards.

Shortly afterwards the Company's outstations at Greenock and Largs started to provide a parallel facility for flying boats. Thereafter Catalinas became the principal type handled at Greenock. The first of many which were to be converted for the RAF arrived on 1st February 1941 and by 1943 Largs was acting as a reception base for all Lease-Lend flying boats crossing the Atlantic. The aircraft were moored on the leeward side of the island of Great Cumbrae and the crews were ferried to and fro by launch. Scottish Aviation's Largs operation finally closed at the end of November 1944 when the reception point for flying boats was transferred to Wig Bay and consequently passed out of the Company's control, although storm damage to the aircraft at their new location shortly afterwards produced the suggestion that Largs should at least be kept open as an alternative terminal.

The Consolidated Catalina PBY flying boat.

The Atlantic air bridge - ferry aircraft on the north apron.

The principal American types on which the Company worked at Prestwick were Liberators and Dakotas, Mitchells, Hudsons and Flying Fortresses. Often as many as between thirty and fifty aircraft would arrive overnight and contemporary photographs of the crowded dispersal areas are testimony to the scale of the Company's activities at this time.

Acting as the European terminal for the Atlantic air ferry gave Scottish Aviation experience of a great variety of aircraft. Prominent among the British types which achieved great distinction in service were the Avro Lancasters and De Havilland Mosquitoes which were by now being built in Canada.

In January 1945 for example, a heavy programme of work on Liberator Mark VIIIs required many employees to put in a 70-hour week, and although the job was successfully completed, a number subsequently fell ill with pneumonia as a result. It was perhaps at this time that some employees who had been working night-shift, looking for somewhere to snatch some sleep, chose a Liberator bomb bay only to be rudely deposited on the hangar floor and awoken when the next shift opened the bomb bay doors as part of the stipulated sequence of checks on the aircraft.

The 1000th Avro Lancaster bomber to be manufactured in Canada.

Inevitably the fact that working at Scottish Aviation was a reserved occupation produced its share of jibes, such as the suggestion that the Company's initials, SAL, really stood for "Scroungers and Loungers" or alternatively "Safe At Last". The epithets were unwarranted. The first job of work to be done on an aircraft in the Palace was one night when a fitter removed the tail of a Spitfire single-handed so that it could be mounted in a jig for repair. In fact much of the Palace's wall sheeting had been broken in transit from Glasgow and when aircraft work first started in the building the sheeting stopped fourteen feet from the ground and the blast walls had still to be erected. Old oil drums and pails pierced by a pick axe and filled with hot coals were brought in so that nightshift work on the Spitfires could start without delay.

When the main administration block caught fire in 1941 and threatened the adjacent hangars, men bodily lifted an Anson which was stranded with neither engines nor undercarriage, carried it out on their backs and stood it on trestles at a safe distance.

It was not unusual for employees engaged on the Liberator conversions at Prestwick to work two 36-hour shifts in a week. One man had to be ordered home when it was discovered that he had been working continuously for 54 hours. A factory news-sheet established by employee representatives on the Joint Production Committee at Greenock consistently described the craftsmen's tools as the weapons with which he, and she, was taking part in the War. Many SAL employees, however, wished to join the armed forces and contemporary personnel records show that a number of exemplary employees eventually persuaded the Company to give them a bad reference so as to engineer their release from their reserved occupation by declaring them to be troublesome and unsatisfactory.

INTER-DEPT. COMMUNICATION SHEET. SCOTTISH AVIATION LIMITED.

To Mr. Taylor, Staffing. *From* Mr. Broughton.

Date 8.2.44. *Subject*

With reference to the attached application for release, this lad is a very bad type, and I have experienced a great deal of trouble with him, and do not therefore, wish to oppose his application.

F. W. Broughton

Liberator Superintendent.

	Exceptional.	Above Average.	Average.	Below Average.	Poor.
Ability to do assigned work.	Exceptional.	Good. ✓	Satisfactory.	Lacking.	Poor.
Knowledge of work.	Complete.	Good.	Adequate ✓	Meagre.	Poor.
Accuracy.	Superior.	Good. ✓	Satisfactory.	Somewhat Inaccurate.	Careless.
Speed.	Very Fast.	Rapid. ✓	Average.	Slow.	Very Slow.
Attitude towards work.	Enthusiastic.	Very interested.	Satisfactory. ✓	Indifferent.	Not interested.
Ability to learn.	Quick.	Easy. ✓	Average.	Slow.	Dense.
Initiative.	Excellent.	Good. ✓	Sufficient.	Lacking.	Poor.
Reliability.	Highly Trustworthy.	Reliable.	Satisfactory. ✓	Irregular.	Unreliable.
Relations with workers.	Excellent.	Very Good. ✓	Satisfactory.	Doubtful.	Agitated.
Total. 1.					
Conduct.	Excellent.	Very Good. ✓	Satisfactory.	Doubtful.	Troublesome.
Timekeeping.	Excellent.	Very Good. ✓	Satisfactory.	Irregular.	Poor.
Total. 2.		640	180		

1.
2.

Average 74 %   Foreman [signature]

A "good" reference for those who wanted to join the armed forces.

Of the various aircraft types handled at Prestwick during the War perhaps the most significant was the Consolidated B24 Liberator. In equipping these aircraft for an RAF Coastal Command role against the German submarine packs Scottish Aviation made a major contribution to the winning of the Battle of the Atlantic. In the autumn of 1942 the Company devised the package of modifications by which the aircraft was converted into a Very Long Range variant, with a range of up to 2,650 miles. This increased capability was achieved by stripping out the self-sealing bags protecting the fuel tanks in the wings, embodying extra tankage in the forward bomb bay and removing all other dispensable weight, including the top gun turret, bomb winches, auxiliary power unit and much of the armour plating. The other major modification which the Company incorporated was the Leigh Light which enhanced the aircraft's capability to seek out and destroy enemy submarines on the surface at night. Its inventor, Sir Robert Watson Watt, also the father of radar, even came to Prestwick to ensure that the vital importance of this installation was understood by the workforce.

Liberator work also produced its share of anecdotes, as when "Bomber" Harris himself came to observe a rocket launcher installation which had been fitted in the aircraft's bomb bay and which had been electrically wired so that it could not be deployed or fired until the undercarriage was retracted. For the purposes of the demonstration the aircraft was mounted on jacks, but unfortunately the battery which was then connected was too

powerful and the aircraft promptly went into an apparently unstoppable cycle of undercarriage, rocket launcher and radome in turn emerging and disappearing so rapidly and violently that they threatened to shake the aircraft off its jacks until the battery was hastily disconnected.
Liberator work originally took place in the Anson Hangar where the relatively low doors required a squad of men to position themselves in the aft end of the aircraft so that the tail would dip low enough to achieve the necessary clearance. Unfortunately on one occasion someone blew the tea-break whistle in the middle of this manoeuvre and, as the ballast suddenly shifted forward to quench its thirst, the fins went straight through the roof of the hangar.

Nose to tail : Liberators

The task of manoeuvring the aircraft into the hangar was made doubly difficult by the position of the building's internal supports with the result that it had to be skewed and guided by hand round the obstacles. Once as the Foreman walked slowly backwards into the hangar directing these operations he tripped and fell just as the aircraft gathered momentum and rolled on into the hangar. The nose wheel pinned his jacket to the floor and he came within inches of being crushed, an incident in which, as he admitted shortly before he passed out, he had "heard the flutter of the angels' wings".

The workload had always been prone to fluctuation and sudden revisions, however. In June 1943 for instance, the Liberator modification programme, which had been running at thirty per month, was suddenly reduced to twelve and arrangements were made to modify twenty five B25s per month to fill the factory's capacity. There were also fears that as the War moved Eastwards it would draw repair and modification work with it and away from the West of Scotland.

In the event this fear proved to be groundless for although the opening of the Second Front with the invasion of Normandy in June 1944 did not produce the expected peak of repair work or the need which had been anticipated for Mobile Repair Parties to travel to the South coast of England, it did produce an upsurge in Liberator and Fortress conversions. As a result, it was in July 1944 that the numbers employed by Scottish Aviation on its factory activities reached their peak in a combined figure of 5,291 at Prestwick and Greenock and a total of some 6,500 on all the Company's activities. In fact, the volume of Liberator and Fortress work was by now such that excess work was passed over to a plant at Renfrew operated by the American Lockheed Aircraft Corporation, who acted, in this instance, as sub-contractors to Scottish Aviation, an interesting reversal of the relationship which was to exist between the two companies a quarter of a century later when SAL was to undertake manufacture of fuselage panels for Lockheed's C.130 Hercules transport.

The Centre Bay of the Palace housing Spitfires, Liberators and a Flying Fortress.

By 1944 Scottish Aviation had established, as a result of its particular role at Prestwick, an expertise in American aircraft which was unique among British companies. The quality of its work had also established the Company's reputation. Much of the Flying Fortress work which was allotted to it was experimental and classified secret. On Dakotas, the Company was selected to convert an aircraft into a luxury air caravan and mobile headquarters for General Maitland Wilson.

Scottish Aviation was also responsible for any major servicing on the Liberator named "Commando" which was used by Winston Churchill until September 1943. The work in this case was done at Lyneham by a mobile repair party from Prestwick who used to travel in their own Liberator all over Britain and Northern Ireland to carry out B24 repairs and on occasions even stood in for the scheduled RAF service to Iceland. One of the memorable by-products of this working party activity was an urgent, and presumably mis-spelt, cable from RAF Oxford demanding to know, "when may we expect your whoring party?"

The expertise which the Company had accrued on American types meant that there was considerable disappointment when the lack of free capacity at Prestwick in June 1944 resulted in an aircraft of the new Douglas type, the DC.4, having to be allotted to Armstrong Whitworth rather than to Scottish Aviation for conversion into the Prime Minister's new personal transport. The Company's concern was not simply at the loss of a prestigious job but, from a business perspective, at the advantage which would be gained by a prospective post-war competitor as a result of this contact with an aircraft of modern American design, an area which had so far been exclusively SAL's province.

The Douglas C.47, the military version of the DC3 Dakota.

It was this competitive advantage which David McIntyre was rightly anxious to pursue to ensure the continuation of the Company's manufacturing interests after the War. The extent of the expansion which had taken place in these activities during this period is reflected in the growth of the Prestwick factory's floor space which had increased from some 29,500 square feet in 1938 to 845,000 square feet in 1945.

Under the stimulus of war Scottish Aviation had come of age. In March 1942 it had added to its existing capability the important new dimension of formal design approvals and, when it sought to join the Society of British Aircraft Constructors later in the same year, it resisted SBAC's initial proposal to assign it the status of a Repair Company only. In 1944 a significant measure of recognition from the industry as a whole was achieved when Scottish Aviation's F.W. Hopkins became Chairman of the Council of the newly formed society for Licensed Aircraft Engineers. The Company had established itself as a major employer and had rapidly developed its experience in many aspects of aircraft engineering and on many types.

At the same time it became increasingly apparent as hostilities drew to a close that the prospects for Scottish Aviation, as a relatively young company and as a sub-contractor in an industry which was clearly going to find itself suddenly with a great deal of excess capacity, were likely to be meagre in the post-war period.

Work on Hurricane rudders finally came to an end in the Spring of 1944. Spitfires continued until June of the following year when they finished abruptly with a loss of nearly 200 jobs. Flying Fortress work ended at about the same time and Sunderlands had disappeared from Greenock by the end of July 1945. At a meeting with the Ministry of Aircraft Production late in August 1944 David McIntyre was notified that there would be insufficient post-war work to spare any orders for Scottish Aviation and the suggestion was made to him that the Company might be advised to concentrate on securing RAF training work of the sort in which it had originally been engaged.

Following a suggestion by Sir Stafford Cripps, consideration was given to the formation of a Scottish Combine of Aircraft Companies with a view to bolstering Scotland's chances of securing a share in post-war aircraft manufacture.

The idea was that the Combine would consist of Scottish Aviation at Prestwick and Greenock together with the Blackburn Aircraft organisation at Dumbarton, in which Blackburn would reduce its interest to that of a minority shareholder. S.M.T., the Scottish Motor Traction Company, which had been involved in aircraft engineering during the war, declined to be involved as it intended to concentrate again on road transport.

In the meantime, as the workload dwindled at Prestwick, SAL embarked on a series of ad hoc projects in an attempt to preserve the engineering skills which it had assembled. These included the manufacture at Greenock of 40,000 steel frames for pre-fabricated houses at a rate of 2,000 per week, and the manufacture of 2,000 plumbing units at Prestwick for the Ministry of Works. Other projects which were considered included the manufacture of a new dry lavatory, designed for aircraft by one of the Company's employees, and the production of an automatic tea-maker called the "Mechanical Maid". Scottish even took out a provisional patent specification for a jet propelled toy boat! The boat was made from press-formed metal sheet and incorporated a tiny copper boiler which heated water and drove it out of a pipe at the stern to achieve propulsion. An application was made to the Board of Trade for an output of 50,000 of these boats but materials for the manufacture of toys for the home market was severely restricted at that time, although unlimited quantities were available for exports. The Company explored the toy's export potential but ultimately it was only a couple of hundred of employees' children who had anything to show for it.

Behind these stop-gaps the Company was trying hard to secure work which would enable it to remain in mainstream aircraft engineering. The extensive work which it had carried out on Dakotas seemed to offer one of the best opportunities when, in the Autumn of 1944, B.O.A.C. made a tentative enquiry regarding the conversion of a fleet of 50 aircraft at Prestwick. At about the same time David McIntyre learnt through the S.B.A.C. that the Indian Government intended to approach Douglas with a similar request. He therefore promptly contacted a Director of the Indian Company which was to operate the aircraft and offered Scottish Aviation's services.

Not content with that, he also initiated a letter in September 1944 to the Douglas Aircraft Corporation proposing that Scottish Aviation should become that Company's agents for the DC.3 on this side of the Atlantic. This audacity was duly rewarded and a licence agreement was concluded in August of the following year.

A similar approach was made to the Pratt and Whitney Engine Company early in 1945, although McIntyre had nurtured the idea for some time and had tried to establish contacts with that Company during his visit to the United States in the previous year. Pratt and Whitney were receptive to the approach, being anxious to have an accredited agency in the U.K. for the overhaul of their engines, but they were, at the same time, reluctant to appoint Scottish Aviation as their official distributors and repairers if the Company were also to become an active airline operator.

David McIntyre, in particular, made no secret of his ambition that SAL should do just that and the principal manufacturing association which he sought to develop for the Company in the post-war period was intimately connected with linking engineering and airline operating activities. This was the notion of an association with the Consolidated Aircraft Corporation in America to enable Scottish Aviation to design and produce a new fuselage for the Liberator to turn it into a civil airliner. As early as March 1943, he was pointing out to his fellow Directors that the Liberator had the most efficient wing available and that it was therefore the most suitable American aircraft for conversion to commercial purposes.

The efficiency of the Liberator's wing made it seem the prime contender for conversion to a civil role after the war.

As early as June 1943 the Company had put proposals to the Ministry of Aircraft Production to build a new fuselage and so utilise surplus wings, but M.A.P. had turned the suggestion down, ostensibly on the grounds of licence difficulties with Consolidated, although one of its Vice-Presidents had already indicated to McIntyre that there would be no difficulty in obtaining his Company's permission for the proposal. This permission was formally confirmed in January 1944. In August, Scottish Aviation was testing M.A.P.'s reaction to the idea that Consolidated should acquire a shareholding in SAL. The reaction was cool. The Ministry indicated that while there was no objection to the transfer of a small nominal shareholding, a greater one would be likely to prove harmful to the Company's relationship with Government Departments.
In the meantime, discussions were taking place with both Rolls Royce and the Bristol Aeroplane Company with regard to engines to power the proposed Liberator conversion. SAL's preference was for the Rolls Royce engine and in February 1945 secret negotiations were taking place between Scottish and Rolls Royce for the experimental installation of gas turbine engines on the Liberator. A stress analysis had been carried out by the Design Office at Prestwick and in March Rolls Royce presented the design study to the Ministry of Aircraft Production, pointing out that the Liberator was the only suitable four-engined aircraft for the installation of the jet-propelled engines owing to the structure of the wings and the position of the retracted undercarriage. Nevertheless, McIntyre was pessimistic that the Ministry would approve the use of an American type for this innovation.

The discussions with Bristol were pursued not so much out of choice but for the pragmatic reason that the Company did not expect to secure Government support for the Liberator conversion unless it were associated with one of the British engine manufacturers. The proposed installation of the Bristol engines was therefore limited to one aircraft only to be used for trans-Atlantic freight and mail transport. In the event, however, the engine proved unsuitable because of the weight effect on payload and range.

Despite these difficulties, the Company decided in October 1944 to proceed to place an order on itself for the manufacture of a prototype fuselage as quickly as possible. Interestingly the name which was selected for this Liberator conversion, and which was registered in October 1945, was "Concord."

By March 1945 enquiries about Scottish Aviation's Concord had been received from both Holland and Iceland, for whom the Company was later to convert Catalinas, but the project remained highly speculative, involving as it did many factors beyond the Company's control, such as the continuing absence of official acceptance of Prestwick as a permanent International Airport, the obstacles involved in the purchase of second-hand American aircraft and the Government's protective refusal to permit British operators to use a converted American aircraft.

These matters were unresolved when the war came to an end both in Europe and in the Far East. In October 1945 the Company was still waiting for the Ministry of Aircraft Production to grant its permission to proceed and was growing increasingly anxious that, unless this authority were forthcoming soon, then the aircraft would have become obsolete.

Already the proposed conversion had lost some of its prospective advantages. It now seemed that while it could compete with the Douglas DC.4, it could not in any way be expected to be competitive against the Constellation and other civil aircraft which were entering production in America.

Scottish Aviation's frustrations at this time were by no means confined to its manufacturing endeavours. There were similar difficulties to be faced in pursuing its aspiration to become an airline operator and these derived in large part, as did the aspiration itself, from the Company's experience in the third major branch of its early activities, that of airport operation.

Prestwick Airport shortly after the War

Chapter 3

"TO THE THRESHOLD OF EVERY MAN'S DOOR"

Airport Operation

Since that first pathfinding Hudson in November 1940 had only landed at Prestwick as a result of losing its way and overshooting Aldergrove, it might be supposed that Scottish Aviation took on the role of airport operator by accident rather than by design. As early as 1935, however, David McIntyre was known to declare that there would be an international airport at Prestwick within ten years and it is clear that from early on he and the Hamiltons saw airport operation as one of the principal branches of the aviation industry which the Company should be seeking to establish in Scotland. More surprising perhaps is the fact that before 1940 the focus of the Company's aspirations to develop this aspect of its business lay not at Prestwick but at Grangemouth.

Grangemouth Airport, which has now totally disappeared under the oil refinery and associated industrial development, was built by Scottish Aviation in 1939 but, like Prestwick, its origins lay in the rapid expansion of RAF training activities which had begun some years before.

In the middle of 1936 Scottish Aviation learned of the possibility of further Air Ministry contracts for Training Schools and promptly began to investigate areas which might be suitable for the construction of aerodromes so that it would be in a position to tender for these contracts at short notice. On the basis of terrain and weather conditions Port Patrick in Galloway, the area between Annan and Carlisle and Buchan in the North of Scotland were selected as the initial target areas and were surveyed for this purpose.

At the same time the Company explored the prospect of establishing itself on existing airfields. Discussions took place with the owner of Aberdeen Airport with a view to a collaborative venture to set up an Elementary Flying Training School there and plans were laid to obtain an option on the site of Drem Aerodrome some 18 miles east of Edinburgh. The Air Ministry's reaction, however, was not encouraging. By the end of 1936 it had turned down the Aberdeen proposal and in January 1937 it indicated that it did not intend to establish a Volunteer Reserve Training Centre at Edinburgh.

It is a measure of the rapidly deteriorating foreign affairs situation during 1937 that within nine months it had changed its mind and had decided that an Edinburgh Centre was necessary as part of a general extension of the RAF Volunteer Reserve Scheme.

By this time the Company had approached the controlling authorities of a number of airfields and its preferred choice as the centre for Edinburgh was Macmerry, several miles closer to the city than Drem, but the Air Ministry then stated that it was its intention to negotiate directly with the aerodrome's controlling authorities, the Town Council, before inviting tenders simply to cover the provision of the Flying training.

Scottish Aviation Limited

request the honour of your company on the occasion of the

Empire Air Day Flying Display,

which is being held at Prestwick Aerodrome

on Saturday, 21st May, 1938,

between the hours of 2.30 and 5.30 p.m.

This card is required for admission to the Aerodrome and should be retained throughout the afternoon for admission to Refreshment Rooms reserved for private guests.

Note.—Empire Air Day is 28th May, but it is being held one week earlier at Prestwick in order to avoid disturbance to the British Amateur Golf Championship Finals at Troon on 28th May.

This was a potentially serious set-back to the Company's plans which aimed not simply to set up another training centre but to develop a site which could eventually become the civil airport for Scotland's capital city. This proposition had already been put to the Edinburgh Town Council which had indicated that although it was unwilling to purchase the aerodrome outright for the city it would be prepared to loan finance for the extension and development of civil airport facilities. Other civic authorities were also recognising the importance of having air transport facilities and both the Bute and Huntley Councils had already approached Scottish Aviation to conduct surveys for them to identify suitable airfield sites. In December 1937 therefore the Edinburgh Town Council advised the Air Ministry that it had authorised SAL to negotiate on its behalf regarding Macmerry.

The next blow came only a month later when the Ministry declared Macmerry unsuitable to be a Volunteer Reserve Centre. The Company switched its attentions back to Drem and an adjacent site at Gosford. It was fortunate that it did so, for in the Spring of 1938 the Air Ministry suddenly revealed that it was actively considering the acquisition of Prestwick to establish its own flying training school. This was the first of a series of threats to the Company's tenure of the airfield which it had created. In an attempt to avert a forced sale of Prestwick, the Company suggested Drem as an alternative and the Ministry duly purchased Drem and Gosford. As these were no longer available for development as Edinburgh's civil airport, SAL again pressed the case for Macmerry.

In fact, Macmerry was not ideally suited to the role. Since it had first been considered for Volunteer Reserve Flying the Air Ministry had twice revised its grading requirements and in December 1938 had stipulated that a gradient of 1 in 100 rather than 1 in 50 would be necessary and would definitely be the standard required for first-class civil airports. The prohibitive costs of conforming to these new requirements, together with the Civil Aviation Department's refusal to guarantee that these minima would enable Macmerry to be recognised as the Edinburgh airport for any length of time, brought SAL's interest in the site to an end and Edinburgh Corporation was duly advised that the Company could not recommend it to proceed.

The Town Council was disappointed but keen to consider an agreement with Scottish Aviation for the provision of airport facilities, if the Company found itself able to suggest an alternative site. The Company for its part was already seeking to do precisely that and in July 1938 had obtained a report on the suitability of a location at Grangemouth.

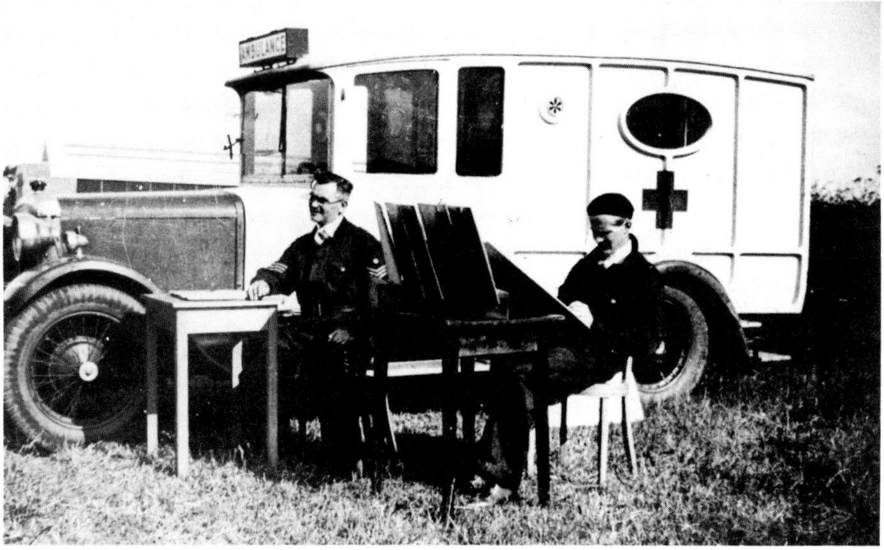

Prestwick Airport emergency services before (1941)

Again the intention was that the new aerodrome should serve both as the Volunteer Reserve Training Centre for Edinburgh District and as a civil airport, but in the latter respect the Company's aim was now to pick up the recommendation made by the Maybury Committee, which had been appointed in 1937 to consider the development of civil aviation in the U.K., and to create a single airport to serve both Edinburgh and Glasgow.

......................and after (1944)

In March 1939 the Company suggested to both City Corporations that they make the modest contribution of at least £750 each towards the operating costs of the new Central Scotland Airport and a further contribution to the provision of luxury airport equipment such as night lighting and hard runways. Edinburgh was prepared to discuss the proposal but Glasgow remained committed to its own airport at Renfrew. The opportunity which Scottish Aviation had presented for the rational development of a single airport for Central Scotland was consequently lost.

But none of this was apparent in the Spring of 1939 and the construction of the new airport progressed rapidly on a 521-acre site which was nearly three times the size of the aerodrome at Prestwick at that time. Work began on February 7th and was completed in record time. With Scottish Aviation's winged tiger's head insignia prominently displayed on the facade of the terminal building and in terrazzo tiling on the floors, Central Scotland Airport was formally opened by Air Marshall Viscount Trenchard on July 1st, 1939. Eight hundred bottles of champagne were consumed, and there was good cause for celebration, because the aerodrome had in fact become operational two months earlier. The two schools operated by Scottish Aviation at Grangemouth, No 35 Elementary and Reserve Flying Training School and No 10 Civilian Air Navigation School, launched their first courses on May 1st and June 12th respectively.

But if it seemed at this stage as if Prestwick might be eclipsed by Grangemouth as the focal point for the Company's activities, the prospect was short-lived. The circumstances of war, which were to prove such a powerful stimulant to the development of the one, had precisely the opposite effect on the other.

The official opening of Central Scotland Airport, Grangemouth, 1st July, 1939.

Grangemouth's proximity to both Edinburgh and Glasgow had been the basis of the Company's ambitions for its development. But this location was now to be its undoing, for with the outbreak of hostilities bombing raids on the two cities not only interrupted the flying training programmes but made Grangemouth a tactically important base for defending fighter squadrons. Ironically, it was Hamilton and McIntyre's own former unit, 602 (City of Glasgow) Squadron, which dispersed a flight of Spitfires to Grangemouth daily during September 1939 from its base at Abbotsinch, although there was some consolation in the fact that 602 was involved in shooting down the first German bomber of the war over Britain.

The airfield was also used by 603 (City of Edinburgh) Squadron and by Fleet Air Arm Walrus flying boats. Scottish Aviation could pride itself in having made another contribution to the war effort, but it came no closer to realising its vision of a Central Scotland Airport than a single operator, North-East Airlines, which used Grangemouth briefly while flying Dragon Rapides on an East coast route between London and the Shetland Islands.

35 ERFTS closed in September 1939 following the cancellation of the major part of its training programmes. Although 10 CANS continued at Grangemouth until December, when its activities were absorbed into the Air Observer School at Prestwick, the Scottish Aviation Board was already seeking to release the assets which were now tied up in the Central Scotland Airport by suggesting that the Air Ministry should acquire it. The Ministry was at first, perhaps naturally since its fighter squadrons were already using the site, slow to respond and when the sale was eventually completed in 1941 the Company received £10,000 less than the £160,000 which Grangemouth had originally cost to purchase and develop.

Even more worrying for Scottish Aviation at the end of 1939 was the fact that the Air Ministry's reluctance to acquire Grangemouth was matched by its apparent desire to requisition Prestwick, where the Company's training fleet was expanding to over 120 aircraft.

Prestwick's status during the early part of the war was a curious one. On the outbreak of hostilities many of Scottish Aviation's personnel, being members of the RAF Reserve, were mobilised. Nominally requisitioned, the aerodrome became RAF Station Prestwick under the command of David McIntyre, not in his capacity of Managing Director but as a Wing Commander, and subsequently Group

Captain, in the RAF Reserve. With all the necessary services already available for visiting Squadrons from the Company's existing organisation without requiring aid from RAF personnel he chose simply to ignore the first requisition notice.

On the other hand he did not miss the opportunity to invoke the powers conferred on him as Station Commander to close the Monkton-Sandyford Road, which separated the two pieces of land which the Company had originally purchased, and then to proceed to extend the airfield to the north of the road into the area now occupied by the factory. Fields which had been allowed to remain in crop and grazing since 1935 were drained and seeded and the Aitkenbrae Farm steading demolished. The overall area of the airfield was increased to 397 acres, more than twice its original size.

It was not until after the aerodrome had demonstrated its strategic importance as the transatlantic bridgehead for aircraft arriving from North America that the Air Ministry felt it necessary to regularise the unusual arrangements at "RAF Prestwick". In the summer of 1941 the site was formally requisitioned and David McIntyre was obliged to hand over the role of Station Commander to an RAF Regular. He did so less than willingly because, although Scottish Aviation had been awarded a Station Contract to continue to provide a wide range of airport services, control of Prestwick's future had now passed out of the Company's hands. McIntyre was already deeply concerned that the new arrangements would prejudice the airport's post-war potential.

DC4's outside the Prestwick Terminal

Not the least of his worries was the creation of a Fighter Command aerodrome at Ayr. This Heathfield site, adjacent to Prestwick, had in fact been Scottish Aviation's original preferred location in 1935 but had not been available at the time because both the Ayr and Prestwick town authorities had the area reserved for their long-term housing plans. The new aerodrome, the first of a novel type in which the emphasis was on dispersal, was an exceptionally fine one, with hard runways, and David McIntyre expected that this would affect the eventual value of the Prestwick airfield.

Ironically, again it was Spitfires from 602 Squadron which became the first occupants at Ayr in April 1941 and in this case insult was added to injury, since on this occasion they were re-positioned from Prestwick itself.

In the event the most troublesome features of Prestwick's neighbour sprang from the sheer proximity of the two airfields. Not only did their circuits overlap but Ayr also fulfilled the role of Armaments Practice Camp and on one occasion at least Beaufighter cannon shells punctured the sheeting of Scottish Aviation's Anson hangar.

David McIntyre might well also have felt aggrieved at the sudden intrusion of external authorities to take control of the Prestwick end of the Transatlantic air ferry. While the concept of ferrying the aircraft by air rather than by sea and so reducing delivery from three months to a few days was the product of Lord Beaverbrook's boldness, the translation of this idea into practice was very largely due to David McIntyre's powers of organisation. It is easy to forget now that in 1940 Transatlantic air crossings by landplanes were still rare and dangerous. There were many who believed that the proposed ferrying service would prove at worst to be impossible or at best one which could only be maintained during the summer months. It was the excellent weather conditions around Prestwick, which the ferry pilots nicknamed their "safety window" coupled with SAL's ability to receive and process the aircraft and their crews, which quickly proved the sceptics wrong. Pilots who were new to the ferry service were said to be advised by those who were already familiar with the journey that all they had to do was to fly east until they came to a hole in the clouds, which was where they should land because that was Prestwick. For them "Safe At Last" was a very welcome definition for SAL.

Initially, by agreement with the British Government, the ferrying was carried out by the Canadian Pacific Railway Air Services with technical assistance provided by BOAC, civilian pilots from Canada and the United States and pilots seconded from Coastal Command.

Scottish Aviation ran the Prestwick end of the service directly from November 1940 until March 1941 when ATFERO, the Atlantic Ferry Organisation, was formed by the Ministry of Aircraft Production to take over and indirectly on until August of the same year when the responsibility was vested in the Royal Air Force and ATFERO became RAF Ferry Command, subsequently Transport Command, under Air Marshall Sir Frederick Bowhill. Nevertheless, the majority of the personnel involved continued to be British, Canadian and American civilians together with Poles, Norwegians, French and other foreign nationals who had volunteered for ferry work.

The task of accommodating and generally looking after these aircrews also fell to Scottish Aviation and particularly to SAL's receptionists at the Terminal Mess. It was a task at which even career diplomats might have baulked. The dangers of their job gave the ferry pilots a particular wildness, as the Russian Midshipman Tolstoy demonstrated when he insisted on flying his aircraft upside down and falling out into the Firth of Clyde. The mixture of nationalities simply made the situation more volatile. There was at least one memorable three-sided fight among the Poles, the French and the Canadians, while the sight of the American pilots walking through Prestwick with silk stockings hanging from their pockets to attract the local girls caused its own tensions among the Allies. If anyone could control them, it was Scottish Aviation's Chief Receptionist, known on both sides of the Atlantic as "Miss Grey", a constant ambassador who was recommended for the OBE by Sir Frederick Bowhill himself after she had persuaded the pilots not to refuse to fly following an accident in which a Liberator carrying ferry crews back to America had crashed into the hills on Arran.

Cathie Grey, Chief Receptionist at the Orangefield

Originally the aircrews had returned to Canada by ship, but this had soon resulted in a bottleneck with badly-needed pilots out of action during the slow sea crossing. In May 1941, a return air ferry was initiated by BOAC using seven Liberators on what became the first scheduled landplane service across the North Atlantic. The aircraft were serviced and eventually crewed by Scottish Aviation. In July of the same year the traffic expanded even further when Air Transport Command began to operate through Prestwick.

By October 1944 Transatlantic crossings were taking place at a rate of over one thousand per month and seven airlines were operating Transatlantic services from Prestwick: BOAC, Trans Canada Airlines, the Royal Canadian Air Force Mail Service, RAF Transport Command, American Airlines Incorporated, Transcontinental and Western Air Incorporated and A B Aerotransport Swedish Airlines. Under its Station Contract with the Air Ministry Scottish Aviation was responsible for the provision of servicing and maintenance facilities for the airport, and therefore covered all these operators, providing services which ranged from aircraft handling to parachute packing and from the maintenance of night flying equipment to the operation of a fulltime fire brigade.

With this volume of traffic and range of operating responsibilities the physical airport facilities at Prestwick developed rapidly during the war years, although, once again, it was only with some difficulty that the Government departments responsible could be persuaded to recognise the particular requirements of its unprecedented role.

David McIntyre had very quickly grasped the extent of the role which Prestwick could play and had recognised the developments which were necessary to enable it to do so. He therefore submitted to J P Bickell, Director of ATFERO at the Ministry of Aircraft Production, the Company's designs for a main runway 2,200 yards in length and 100 yards wide, arguing that nothing short of this would be adequate for the receipt of the Transatlantic traffic. At the time the Ministry's standard for runways was only some 1,000 yards in length and 50 yards wide and the Scottish Aviation proposal was not well received. Bickell, however, who was a Canadian, was won over by the logic of the Company's case and in the end offered his resignation unless SAL's recommendations were accepted. Before the end of the war Prestwick was handling as many as 70 aircraft movements an hour and the extraordinary runway length had entirely justified itself. Indeed,

"Safe at Last", Transatlantic arrivals.

that first hard runway, constructed in 1941, was followed by a second of 1500 yards in 1942.

One of the ready advantages which Prestwick had to offer, when the Transatlantic air ferry commenced was the existence of the necessary radio communications facilities which had already been established in connection with Scottish Aviation's air observer navigator and wireless operator training activities. It was not long before the Transatlantic Area Centre for Air Traffic Control was moved from Gloucester to Prestwick and established in Redbrae, another of the large private houses which had been drawn into the war effort.

It should be said that in the first instance Scottish Aviation had certain reservations about becoming involved in some of the activities associated with its responsibility for airport operation under the Station Contract. Catering, for instance, was a far cry from the Company's original aims, but, as the Board decided in September 1941, although it was not directly interested in the terminal messing arrangements, it was preferable to accept rather than to see the work allocated to external contractors.

This desire to protect its ability to control the destiny of the airport which it had created was a natural and very evident concern of the Company's. After the Air Transport Auxiliary had established its first base outside London at Prestwick, the SAL Board monitored its development closely because it was concerned that ATA might extend its role from the onward delivery of ferried aircraft to the provision of servicing and other airport facilities.

Some of the results of the Company's insistence on being involved in every aspect of the airport's development were more curious. By 1944, for instance, Scottish Aviation had its own orchestra providing music in the Terminal mess on Sunday evenings, but even when it was agreed that the cost of this ensemble was disproportionate to other publicity expenditures, it was decided that rather than drop this feature altogether, members of the Reception Staff should be asked to provide the music instead.

The importance which airport management had come to occupy in the Company's thinking by this time is readily understood when one reflects on how far the airfield at Prestwick had come during the first ten years of its existence. What had begun as a grass strip used by a handful of Tiger Moths had become in that short space of time nothing short of the busiest international airport in the world. 37,000 Transatlantic ferry flights alone passed through Prestwick during the war apart from other military movements. At the peak of this traffic in August 1944 the airport handled 7,847 flights in a single month. The entire American 82nd Airborne Division went through during one weekend.

The focal point of these activities was the Orangefield, subsequently the Airport, Hotel which served as the Terminal. Orangefield House had been built around 1690 on the site of what had previously been Monkton Castle, and had been purchased by a Merchant Venturer called McCrea who had once been a poor messenger boy in Ayr before he went to sea, joined the East India Company and eventually became Governor of Madras, accumulating a fortune in the process. The poet Robert Burns, a later visitor to the house, described it as a "pleasant spot in sandy wilds", words which were carved in the sandstone above the entrance.

By 1945 the sandy wilds were long gone. Scottish Aviation had originally used part of the Hotel to accommodate pupils of the flying school and went on to add annexes, squash courts and a swimming pool. The Company's civil engineering department also built the control tower which stood on the roof of the hotel, perched in fact on the four cement-filled chimney stacks of the original house where, contrary to the expectations of many who saw it, it withstood the worst that even Ayrshire gales could produce until the whole building was demolished to be superseded by the new terminal in 1964.

Orangefield House, the Airport Hotel and Terminal

In 1945, however, Prestwick was the best-equipped airport on this side of the Atlantic and the Orangefield terminal was the finest in the country. At a time when both Heathrow and Northolt, the aerodromes which served London, were little more than grass fields with a few huts, Prestwick was already a modern airport at which aircraft could taxi right up to the terminal and passengers could find not only food, drink and accommodation but the full range of associated services, including a bank, a post office and a hairdresser's.

Many famous individuals passed through Prestwick during the War including Eleanor Roosevelt, wife of the American President, comedian Bob Hope, Yehudi Menuhin who gave a violin recital in the Scottish Aviation canteen and stilled its usual clamour, and the actor Clark Gable who did not make himself popular with the girls in uniform at Prestick by demanding in a loud voice, "What are these women doing? Don't they know this is a man's War?"

Reception Hall, Prestwick Airport c.1947

A less-publicised visitor was a Petlyakov-8 four-engined Russian bomber which arrived at Prestwick in May 1942 for repair before proceeding on its journey to carry the Soviet Foreign Minister, Molotov, from Moscow to Washington to meet Roosevelt. The Russian aircrew made their mark with the girls in the factory by giving away their red star cap badges.

David McIntyre recognised early on the epoch-making significance of the Hudson which had landed on 29 November 1940 and every year thereafter he celebrated the occasion with an Anniversary Transatlantic Dinner. It is a measure of the international recognition which Prestwick had secured that the fourth of these dinners in November 1944 was attended by the Air Attaches of China, Czechoslovakia, Yugoslavia, Greece, Poland and Russia.

It is small wonder therefore that Scottish Aviation considered itself to have, in the airport, an asset of prime importance for its post-war prosperity, but to appreciate fully the importance of the airport to the Company's objective of establishing the aviation industry in all its branches in Scotland it is also necessary to understand David McIntyre's personal views on the nature and potential of aviation.

As early as July 1942 he was expounding these views in detail to the editors of the Scottish Press, also regular invitees to his Transatlantic Dinners. He was convinced that aviation would become not only one of the world's major industries and sources of employment but its prime medium of trade. He was determined that Scotland should grasp the opportunities which this presented.

> "The northern industrial area, which has shown such enterprise and effectiveness in the shipping and heavy engineering industries, and which has now become so dangerously dependent on them, will inevitably have to direct its energies into this new channel of world trade, and it is to be hoped that it will do so in a forthright and timely manner by creating a native interest in overseas airline operation, which is the root cause of all aviation markets".

Prestwick airport was, in his own words, a "foot-hold" in the sphere of international airline traffic. That McIntyre regarded airline operation as the highest branch of aviation and the one which SAL's aircraft manufacturing activities existed to serve, is clear not simply from his notes to the Editors of the Scottish Press but from the views which he expressed regularly to his fellow Scottish Aviation directors.

A war-time view from the Prestwick Control Tower

When the Board Meeting of July 1944 discussed the criticism which had been levelled at the Company for attempting to undertake the three distinct activities of Airport Control, Airline Operation and Aircraft Manufacture, he indicated that he felt the criticism to be fair and suggested that as soon as other aircraft manufacturers could be established in Scotland, perhaps from firms previously engaged in shipbuilding, a large proportion of Scottish Aviation's manufacturing could be handed over to them. In an editorial for the Greenock Factory Newsletter in January 1945 he suggested that the Company intended to make the former Flying Boat base its principal production unit and to develop Prestwick first and foremost as an airport.

It is also evident that the Hamiltons and McIntyre regarded Prestwick as an important national asset for Scotland and Scottish economic life and it was characteristic of them that they were prepared to forego the commercial interests of their own company in order to secure this asset. It was in this connection that Scottish Aviation reached the parting of the ways with its De Havilland shareholders in 1943.

At a Board Meeting in The Waldorf Hotel in London in March the Duke of Hamilton reported that the Secretary of State for Scotland and others were actively interested in the prospect of Prestwick becoming a designated International Airport. Hamilton felt, however, that the Air Ministry and Government officials were suspicious

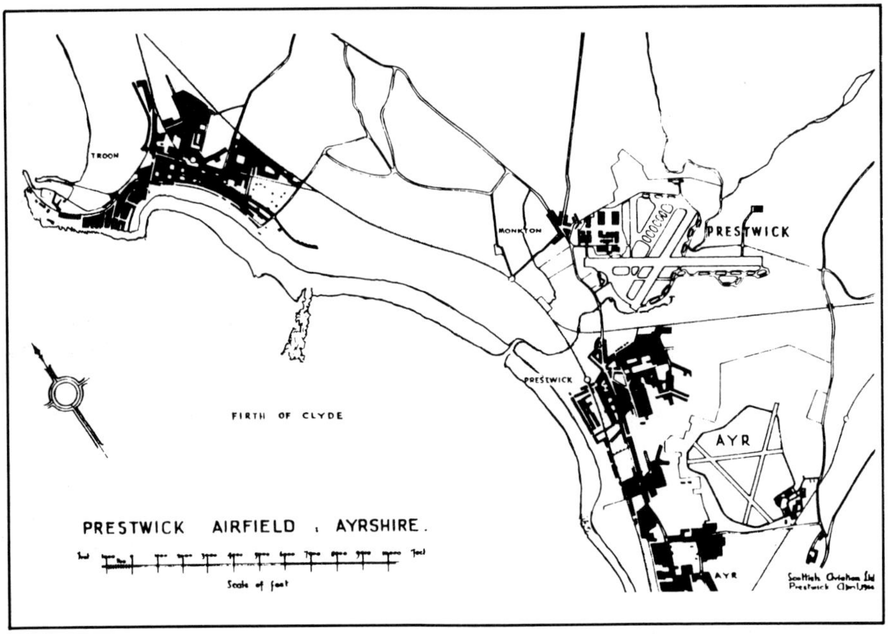

As it was, 1944

that the Company's advocacy of Prestwick was prompted primarily by the profit motive and that this suspicion was retarding the development of the airfield. He sought the Board's approval to write to the Secretary of State to make it clear in official quarters that the Company was prepared to co-operate in the formation of a Public Utility Corporation to run the airport, if and when that was considered more appropriate.

The proposal was a controversial one and although it was supported by McIntyre and the Earl of Selkirk, certain of the other Director-Shareholders, including W E Nixon the Chairman, made it clear that they could not be associated with any statement which suggested that the development of aviation would not be handled most efficiently by private enterprise, even though it was generally agreed that the ownership and operation of an International Airport was unlikely to be commercially economic. Nixon suggested the alternative of offering the airfield for sale to the Air Ministry but this did not resolve the concern that the airport might not be developed in the best interests of Scotland in the post-war period.

Temporarily at least this difference of opinion was set aside by means of a letter stating simply that "The Company desires to assure the Secretary of State for Scotland that on considering the question of aviation in Scotland he could count on the fullest co-operation of Scottish Aviation Limited in arriving at the best decisions in the National interest".

The Scottish Secretary for his part subsequently met the Duke and confirmed his interest in Prestwick by undertaking to forward to the Secretary of State for Air a resolution by the Scottish Council of Industries nominating Prestwick as an International Airport.

In the meantime there was still no declared Government policy on civil aviation and the longer the delay the greater the anxiety within SAL that Scotland's interests were being overlooked.

At the September Board Meeting Hamilton presented a draft letter which he proposed should be sent to the Secretary of State for Scotland indicating that the Company considered that it had advanced its various branches of aviation development as far as they could be taken by a Private Company.

> "We are therefore proposing to alter our Articles of Association to those of a Public Utility Company, paying only a limited return on its Capital and to strengthen the Board by the addition of representative Directors who appreciate the possibilities of World Aviation and who have the welfare of Scotland sincerely at heart".

Recognising that this was likely to be at variance with the policy of the De Havilland interests, Hamilton, who together with McIntyre represented the majority shareholding, had already discussed the matter with the Chairman. Nixon had confirmed that, particularly as Scottish wanted to develop an airline, his Company, which had airline operators among its customers, would wish to be relieved of its shareholding.

David McIntyre's vision of how it should be

An investigation into the best means of transferring the interests of the dissenting shareholders was promptly put in hand. It is indicative of the amicable nature of the divorce that at another meeting in the Waldorf Hotel in November De Havilland intimated that, subject to the agreement of any other retiring parties, it would be prepared to accept 30 shillings per share for its interests, although the apparent book value of the shares at the time was approximately £3. The majority shareholders for their part reiterated that the policy which they wished to pursue was one which contained no wish to make a profit except to invest in the advancement of aviation in Scotland and they offered that, in the event of the Company going into liquidation or the shares being sold, they would hand the difference in price over to the retiring Shareholders.

The close association between De Havilland and Scottish Aviation therefore came to an end after eight years. The Duke of Hamilton took over the role of Chairman and it was illustrative of the Company's re-orientation that the first meeting after the split was held in Edinburgh.

The nature and extent of Scottish Aviation's ambitions for Prestwick Airport were documented in a series of Target Development Plans, the first of which was submitted to the Air Ministry as early as August 1942. The Target Development Plans were nothing if not imaginative, envisaging as they did the progressive extension of the main runway from its already unusually long 2,200 yards to the then unprecedented length of 7,000 yards with a simultaneous doubling of its width to 200 yards. One of the best measures of the radical nature of this scheme is that it planned to push the runway across two of the country's most famous and revered golf courses, Old Prestwick - original home of the Open Championship - and Old, now Royal, Troon. McIntyre's aim was to extend the runway to the sea and to integrate it with slipways to provide a vast flying boat bay, anticipating as he then did that these aircraft would continue to have a major role in Transatlantic air travel and that huge flying boats carrying as many as 1,000 passengers were a real possibility.

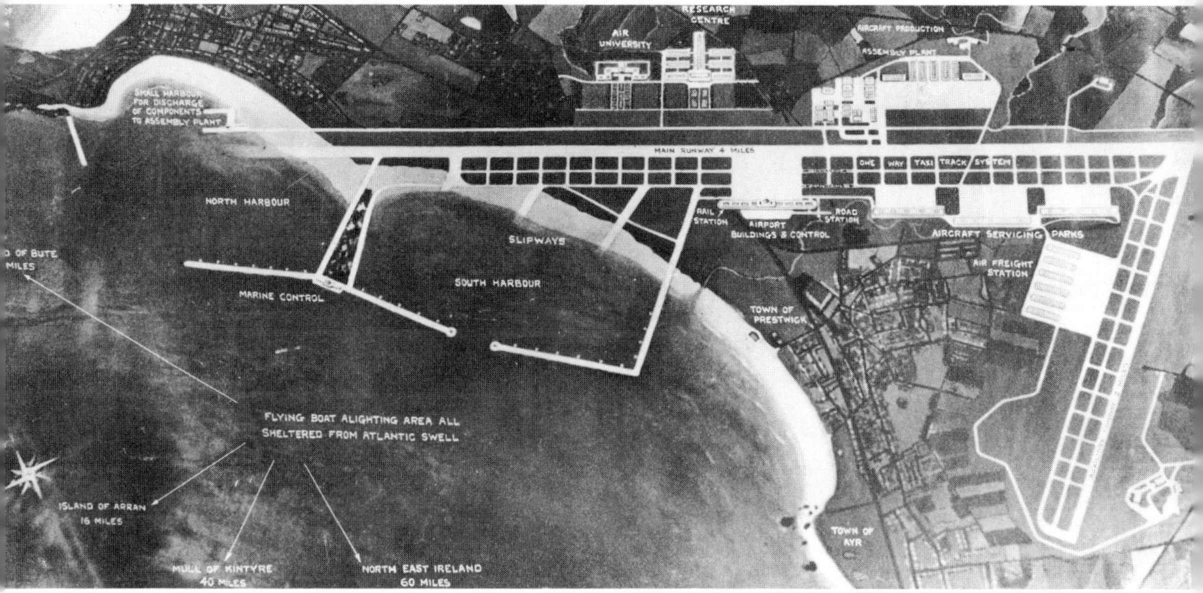

The Target Development Plan, April 1944

The plans also provided for the construction of a new secondary runway to be used in the event of gales from the South West and affording a 1 in 50 glide path from the North East to replace the 1 in 32 descent offered by the existing secondary runway which had the added disadvantage of approaches over residential areas.

The main airport buildings, including control tower, hotel and traffic halls, were to be located in a central position with a one-way system of taxi-ways for incoming and outgoing aircraft respectively. The main road and eventually the railway line were to pass by means of cut and cover directly under the main runway. When in due course the runway was extended and before the present loop road was constructed, the main road passed across the runway and for many years traffic between Prestwick and Monkton had to halt at traffic lights to allow aircraft to take off and land.

McIntyre's vision for the railway was that there should be a high-speed connection between Prestwick and Glasgow by means of monorail - another Scottish invention - and that in due course this should be extended from Glasgow to Edinburgh and from Prestwick South to Carlisle and beyond. His development plans for Prestwick were therefore not simply to turn it into the principal Transatlantic Airport for the UK, but also to revive the earlier aspiration that Scottish Aviation should manage and operate the main internal airport for Scotland, located however not outside the capital or even between Edinburgh and Glasgow but at Prestwick itself.

The principal objective in publishing the Target Development Plans was to draw attention to the fact that since the requisitioning by the Air Ministry in July 1941 a variety of national authorities and agencies had been involved in the lay-out and management of the airfield and its ancilliary buildings and services, often with different objectives and with the result that development had taken place in an ad hoc and inconsistent manner. Typical of these inconsistencies was the fact that, while the Air Ministry Airfields Board sought to acquire the site in 1938 on the grounds of the exceptional suitability of its weather conditions for flying purposes, a report widely circulated by the same Board in January 1944 stated that careful investigation of Prestwick showed that it would not provide an aerodrome "of even local dimensions".

The Company construed this as a primarily political statement, designed to justify the expenditure which was already taking place on the development of Heathrow. Equally its own Target Development Plans were lobbying documents and the Company worked hard to secure support ranging from the Secretary of State for Scotland to the local Town and County Councils, the Clyde Navigation Trust and even the Department of Health for Scotland. This support was readily forthcoming but the Air Ministry and its Works and Buildings Department, to which responsibility for maintenance and construction work on the airport had been passed, were less willing to endorse the proposals.

In putting the plan to the Air Ministry in 1942 the Company argued that Prestwick had already proved its suitability as a Transatlantic terminal and that agreement on the airport's planned development would enable war-time expenditure to be consistent wherever possible with its peace-time use in that capacity. It proved impossible to obtain agreement. The Air Ministry adhered rigidly to the view that it was not empowered to plan beyond the estimated requirements of the year 1945.

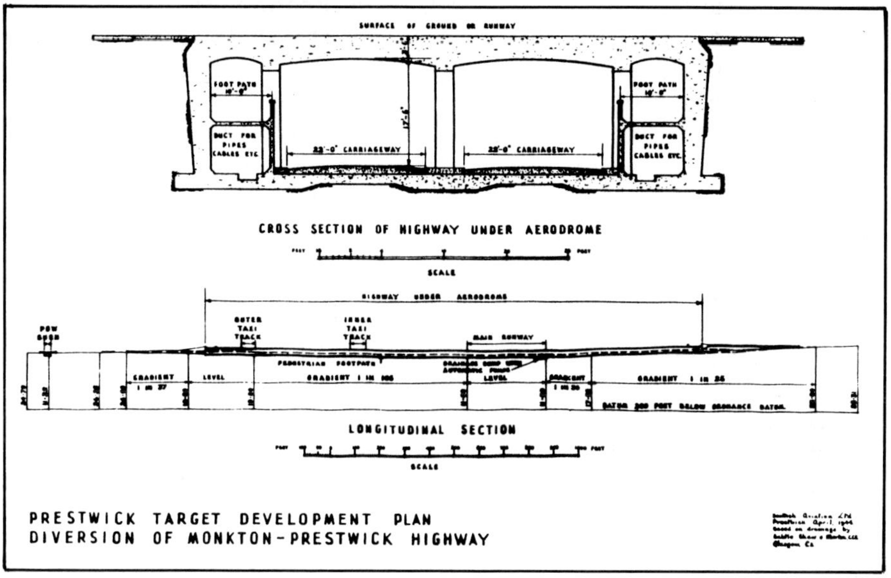

The main road would have passed under the runway

In November 1944, an International Civil Aviation Conference in Chicago had attempted to reconcile different national views regarding the regulation of civil air traffic. The Conference resulted in 22 countries, including the United States but excluding Great Britain, subscribing to what became known as the Five Freedoms of the Air, which essentially allowed aircraft, irrespective of their country of origin, to fly around the world and land at any airport designated for international traffic by the Government of the country concerned. Britain chose to stand off and to pursue bilateral agreements with individual countries for the right to operate commercially into Great Britain. The effect of this and of not designating Prestwick as an International Airport immediately after the war, was to divert traffic away from Prestwick and to stimulate the rapid development of Shannon Airport in Eire. Shannon became a major Transatlantic staging post in Europe and was clearly used by the American Government as a bargaining factor in its negotiations with Britain regarding civil aviation.

Despite its massive historic contribution to the development of Transatlantic flying, the future of

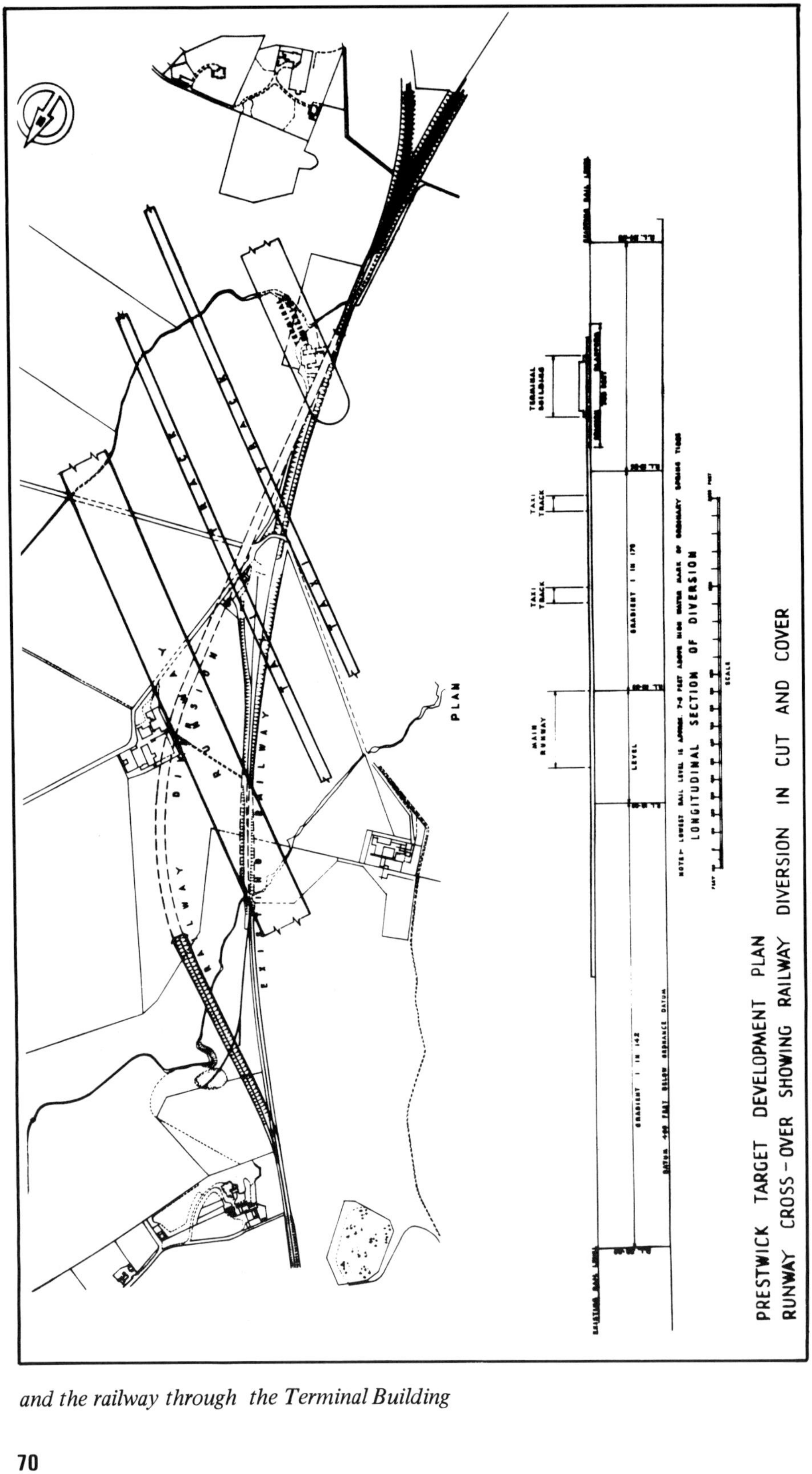

and the railway through the Terminal Building

Prestwick airport remained very much in the balance. It was not until Easter 1945 that a Civil Aviation Debate took place in Parliament. MPs of both major political parties expressed their support for the development of Prestwick, but a statement by the Minister, Sir Stafford Cripps, made the Government's position plain.

> "One cannot afford to have more than one really first-class airport. As the House knows, it has been decided that the Heathrow Aerodrome, which is in course of construction and will cost a very large sum of money, is to be utilised as the main central Transatlantic airport of this country".

But just when the battle to win a place for Prestwick in post-war civil aviation seemed to have been lost, a partial victory was secured when the Under Secretary of State for Civil Aviation announced in May 1945 that the Government proposed to designate it as one of the terminals within the UK for international traffic.

The Burgh of Prestwick for its part chose the same time to bestow its freedom on the Duke of Hamilton and David Fowler McIntyre "For Meritorious Services to Aviation and to Prestwick Airport."

Prestwick's victory was tempered, however, and Scottish Aviation's position rendered once more uncertain by the Ministry of Civil Aviation's subsequent intimation in December 1945 that it intended compulsorily to purchase the Company's airport holding and to assume the position of landlord from 1st January, subsequently amended to 1st April, 1946.

On 19th January 1946, the Scottish Council on Industry called a meeting, which was attended by MPs, Peers, Members of Local Authorities, Chambers of Commerce and Trade Unions, to discuss the future of Prestwick Airport. The meeting passed a unanimous resolution which clearly reflected the Company's own views.

> "That both in the administration of airports and in the operation by the proposed Statutory corporations of air services within Scotland, and direct services between Scotland and the remainder of the United Kingdom and countries overseas, an adequate devolution of the powers of initiative and management should be guaranteed to Scotland.
>
> The Scottish Council on Industry emphasises these points because it believes that Scotland has made

> and can make an effective contribution both to British and world aviation and to the manufacturing aircraft industry, and because, if denied the full opportunity to do so, the structure of economic life in Scotland will be seriously undermined".

Yet while the Company's founders were committed to seeing Prestwick developed in the best interests of Scotland, their vision was at the same time fixed on wider horizons.

There is a story of how on one occasion, Wendy Wood, the prominent Scottish Nationalist politician, arrived at Prestwick to tour the factory and to meet David McIntyre and how he contrived always to be one stop ahead of her throughout her itinerary, an elusiveness which was echoed in his response when she told him, "You must be a Nationalist!" "No," he replied, "an Internationalist."

Among McIntyre's favourite sayings, and an insight into his view of aviation, was the idea that "The air is an ocean that washes over the face of the whole earth."

Group Captain McIntyre and Group Captain the Duke of Hamilton receiving the Freedom of the Burgh of Prestwick 27th June, 1945.

Similarly the menu cards for his Transatlantic Anniversary dinners always bore the inscription

> "An uninterrupted navigable ocean that stretches to the threshold of every man's door, ought not to be neglected as a source of human gratification and advantage."

Underlying his wish to manage and control the development of Prestwick Airport was a more fundamental desire to operate airline services. As Scottish Aviation entered the post-war period it was upon the establishment of this particular branch of aviation that his thoughts and energies were to be concentrated.

Passengers boarding Scottish Airlines Dakota

Chapter 4

"THE WORLD O'ER"
Scottish Airlines

His duties as Managing Director limited David McIntyre's opportunities to take to the air himself but he retained a pilot's passion for flying throughout his life. Although he added little more than one hundred and fifty flying hours during the six years of the war, his log-books show that he made a point of piloting, if only once, virtually all of the aircraft types handled by Scottish Aviation during that time.

Yet it would be too simplistic to ascribe the creation of Scottish Airlines - and the parent company's perseverance with this difficult offspring - simply to McIntyre's enthusiasm for flying. It can more properly be traced to his view that of the various branches of the civil aviation industry the primary one was airline operation, since it was this which was "the root cause of all aviation markets" and gave rise to the demand for aircraft manufacturing and for all the associated aircraft and airport services.

The Liberator used for the Scottish Airlines' North Atlantic route survey flight.

His views on this subject owed much to the insights which he had gained into Transatlantic air traffic as a result of Prestwick's unique role during the war. When, early in 1945 these views were published at length in a pamphlet entitled "North Atlantic: A Scottish Aviation Survey," there were more aircraft on regular daily scheduled flights between Prestwick and Washington than there were trains between Glasgow and London.

> "The war has forced us", he wrote, "into a radically new concept of human relationships, world travel and world trade. Nations linked by direct air service are only hours apart. Cargoes committed to the air carrier are being delivered on the other side of the globe within a couple of days. At the war's end, commerce will fly where the supply planes fly this morning. No spot on earth will be more than 40 hours away from your local airport. In a word, air transport is transforming the business of living."

The conclusions which he drew from his analysis were these. First, that there would be no "favoured nations" in post war aviation and that Britain could not expect to enjoy in this sphere of transport the predominance which it had long held in shipping. Secondly, that the scale of post war commercial aviation would be too big for a monopoly, any "single chosen instrument," to handle. Thirdly, that 90% of the world's population and over 90% of its industry and manufacturing lay in the Northern Hemisphere, thus making the North Atlantic potentially the single most important air route in the world and one that Britain's pre-war civil aviation services, which were largely based on reaching the Empire, had largely ignored. Fourthly, that America was far better organised than Great Britain to seize these opportunities, both by virtue of the competitive civil airline environment which had already been developed in the United States before the war and because the US Air Transport Command had contracted most of its work to American civil airlines which would therefore be in a position to make an immediate transition from war-time to peace-time operation on key routes. Finally, that the competition which would exist in this market would drive fares down and result in a huge expansion of commercial flying.

> "The question of the moment is not whether travellers, shippers and users of air mail should have the benefits of direct air service on the North Atlantic at the end of the war, but how quickly such a service can be organised and who is to provide it."

McIntyre's answer to the question was of course that Scottish Aviation, with the experience which it had acquired at Prestwick during the war, could and should provide this service. The development of overseas and internal airlines operated by SAL was therefore the primary element in his Target Development Plan for Prestwick.

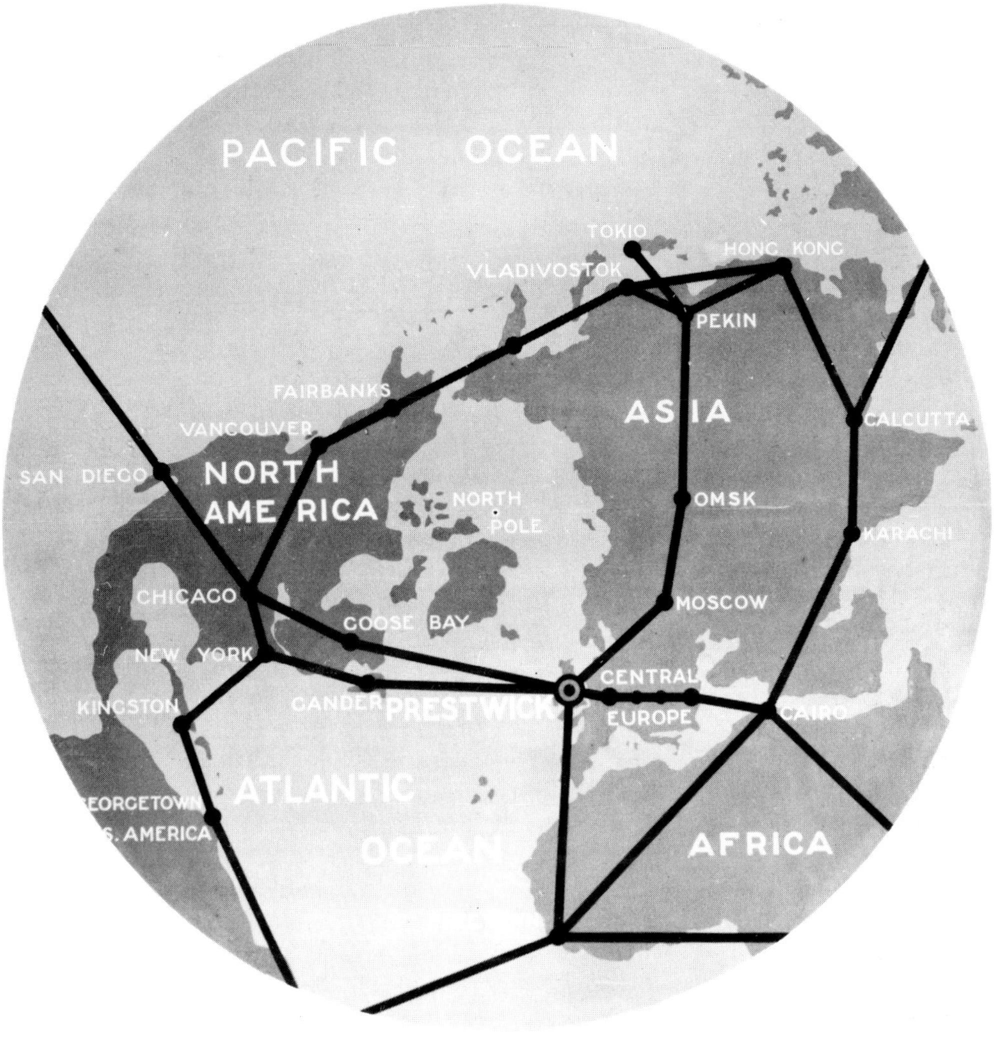

Scottish Airlines' proposed network of scheduled international routes.

The overseas routes which the Company proposed to operate included a night service to and from New York at £80 return for sleeper passengers plus first-class mail and freight with a day service to and from Canada and the United States via Iceland and Goose Bay for passengers and second-class goods. In addition there were to be direct connections between Prestwick and the European cities of

Paris, Amsterdam, Hamburg, Berlin, Oslo, Stockholm and Lisbon, and, most ambitious of all, a global route around the northern Hemisphere linking Moscow, Peking, Vladivostock, Alaska and Vancouver with loops serving Central Europe, the Middle and Far East by way of Vienna, Belgrade, Istanbul, Cairo, Karachi, Calcutta, Hong Kong and Tokyo. As usual, David McIntyre's plans were laid on the grand scale.

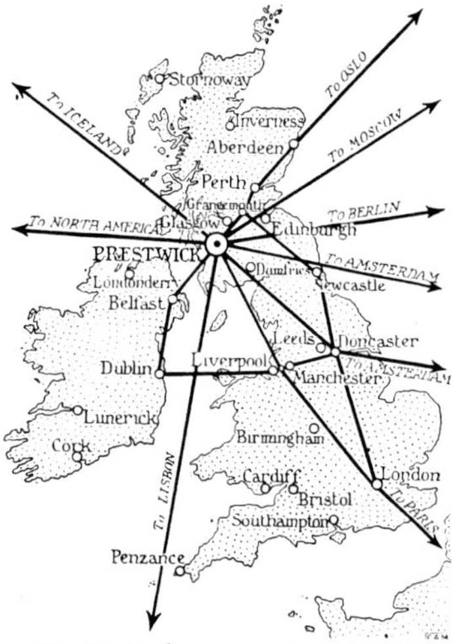

The proposed domestic and European routes

For the internal British routes McIntyre proposed a shuttle service based on Prestwick connecting with London, Liverpool, Manchester, Doncaster and Newcastle, Grangemouth, Perth and Aberdeen, Dublin and Belfast. A novel integral feature of the proposed internal service was the configuration of a number of Dakotas as customs clearance aircraft so that Transatlantic passengers would be able to complete customs and immigration formalities during the course of a short onward flight from Prestwick to their destination within the United Kingdom, thereby avoiding any waste of "door-to-door" time on the ground.

This then was the plan which sprang from a vision that was not simply commercial but reflected David McIntyre's deeply held belief that air transport would prove to be the most potent civilising influence of the Twentieth Century by virtue of its ability to transcend previously inaccessible frontiers and to stimulate trade, employment and prosperity, bestowing "an honest, happy and independent livelihood to more people of all Nations".

This same vision lay behind the motto, "The World O'er", which Scottish Aviation adopted to accompany the Coat of Arms which it devised in May 1945. The quotation was suggested by the Earl of Selkirk following Robert Burns' Poem, "A Man's a Man": "then let us pray that come it may
as come it will for a' that,
that man to man the world o'er
shall brothers be for a' that."

In fact at one stage it was intended that the Coat of Arms and motto should be applied specifically to Scottish Airlines but in the end they were adopted by Scottish Aviation itself.
"The World O'er" was an appropriate measure of the Company's ambitions for its airline operation and it was equally characteristic that in his original design for the Coat of Arms, McIntyre should fall foul of the Lord Lyon King of Arms, custodian of Scottish heraldry, by seeking to incorporate a crown reserved for royalty rather than the ducal crown which could be permitted by virtue of the Duke of Hamilton's association with the Company.

Fokker F.XX11 back in civvies!

The actual beginnings of the Company's involvement in airline operation were rather more modest and lay in its formation in 1939 of a subsidiary known as Night Air Transport through which it was intended to provide a fast night airmail service between Scotland and London operating out of Grangemouth. It was for this purpose that the Fokker airliners, eventually employed on training as flying classrooms, had originally been purchased. The railway companies, however, which had themselves gone into the field of airline operation in 1934 with the formation of Railway Air Services Limited, opposed the Company's application to provide this service and the outbreak of war prevented an appeal being lodged with the Airline Licensing Authority.

The opportunity to pursue its ambitions as an airline operator was effectively suspended during the war but the impulse quickly re-emerged with the prospect of peace and the Company applied as early as May 1944 for the right to operate Transatlantic and European routes.

The first of the various difficulties which the Company encountered in seeking to establish its post-war airline operation was the selection of a name. Curiously history repeated itself and once again the Company found that its preferred designation was considered to be too similar to that of another organisation already registered and initially the name Scottish Airlines was declared to be unavailable.

The Company sought to purchase surplus American types

A more fundamental problem was the difficulty of obtaining suitable aircraft. With a view to capitalising on the unique experience which it had acquired on American types and to implement its designs for their conversion the Company sought to purchase Liberators and Dakotas. When it became apparent that it would be difficult to obtain these from British sources, the Company approached the American Foreign Economic Administration. But although their longstanding acquaintance with Scottish Aviation prompted the American Authorities to waive their normal requirement that submissions to buy surplus aircraft could only be made through the British Government, the procurement process was still protracted and it was not until November 1945 that the Company was able to purchase aircraft.

"There are still seats available for this flight!"

The next difficulty arose from the fact that the purchases proved to be more expensive than had originally been anticipated. The Company had expected to acquire four out of a total of nine surplus Liberators which were lying at Prestwick for £2,000 each. Eventually it was able to obtain three of these aircraft from the Ministry of Aircraft Production but at a cost of £7,500. Similarly, foreign exchange restrictions prevented the Company from paying in dollars for three surplus Dakotas at £6,200 each and the deal which was finally concluded with the Foreign Economic Administration gave the Americans the option of taking the aircraft back after six months and so effectively entitled them to the conversion work which the Company had carried out at a further cost of £6,000 apiece.

The scale of David McIntyre's plans for Scottish Airlines compounded this problem of costs. In April 1945 he had made a formal application through the American Embassy for the purchase of seven Liberators and twenty C47 Dakotas. In the event the initial cost of the three Liberators plus the conversion costs of these three aircraft, the three Dakotas and the Company's Fokker FXXII alone came to a total of nearly £50,000. At the same time McIntyre was urging the Board to place orders for three Lockheed Constellations, the latest in American civil airliners, just as years later he was to press for the new De Havilland Comet, despite the fact that the capital cost of the Constellations would have amounted to £500,000.

"With these", he wrote, as he lobbied Press and Members of Parliament alike for support for an independent Scottish airline, "we could give every other European country a run for its money, and perhaps even take the lead". In fact, by 1949 Scottish Airlines was one of the biggest independent operators in the world.

The principal obstacle to its development, however, was political. In 1946 the Government nationalised Britain's scheduled airline services, bestowing a monopoly on three State Corporations - B.O.A.C., B.E.A. and British South American Airways.

Scottish Airlines had planned to commence its own scheduled services on the Prestwick-Belfast route using its Fokker G-AFZP, from 1 January 1946, but in the event approval from the Civil Aviation Department took time to secure and the service was eventually initiated on 28 January using the Dakota G-AGWS, "Whisky Sugar". The aircraft was configured for 21 passengers and the journey took forty minutes, which, as the Company pointed out, compared very favourably with a surface journey which was then taking 25 hours. The fare was thirty shillings, although the Company hoped to reduce that and to increase the frequency of the service as soon as fuel restrictions were lifted and its aircraft ceased to be limited to 60 hours flying per month.

SCOTTISH AIRLINES
PRESTWICK AIRPORT, AYRSHIRE
Operated by Scottish Aviation Ltd.

Passenger Ticket | No. 3429

ISSUED AT | DATE OF ISSUE 7/4/46

FROM PRESTWICK
TO BELFAST AIRPORT

NAME
FARE £1 10/- DATE OF FLIGHT 15/6/46
FLIGHT No. 3 TICKET ISSUED BY
COACH 2/6 TAXI 2/-

Time of Departure refers to departure of aircraft from tarmac. Please enquire regarding latest time for reporting at office or airport.

When nationalisation denied Scottish Airlines the opportunity to operate its own scheduled services after August 1946, the Company's capacity was transferred to charter work. The Prestwick-Belfast route, for instance, continued to be operated under contract to British European Airways.

It was also under contract to BEA that Scottish Airlines' Dakotas took on the schedules previously operated by Railway Air Services and Scottish Airways between London and Aberdeen, Glasgow and Prestwick. The arrangement only lasted for about six months and at the end of July 1947 BEA took over these routes itself.

This was a pattern which was to be repeated in several of Scottish Airlines' early charter ventures as it opened routes on behalf of a number of European operators who did not have the equipment or expertise immediately available to do so for themselves in the aftermath of the War.

In August 1946 the Prestwick-Amsterdam route was opened for KLM, Royal Dutch Airlines, and the success of the schedule was evident from the fact that after the contract was terminated at the end of March 1947, KLM recommenced the service using its own aircraft from the middle of May. Similarly, during most of 1947 Scottish Airlines flew a scheduled route between Prestwick and Paris on behalf of Air France and regular charters from Prestwick to

Brussels, via Manchester, for the Belgian COBETA (Compagnie Belge de Transports Aeriens).

A longer running association was established with Iceland, which had already expressed interest in Scottish Aviation's designs for converting Liberators and Catalinas for commercial operation. On 27 May 1946 Scottish Airlines opened routes on behalf of Flugfelag Islands h.f. These were a weekly service linking Prestwick and Reykjavik and a twice weekly service connecting Reykjavik, Prestwick and Copenhagen, initially with Liberators and Dakotas respectively but within a couple of months the traffic was such that the larger 24-seat Liberators were applied to all the schedules and the frequency of the Prestwick-Reykjavik service was increased to thrice weekly. Flugfelag eventually took over the routes with its own equipment in April 1948.

A Liberator on the Iceland routes

Other charter work carried out by Scottish Airlines during the early period of its history included fortnightly flights from Prestwick to Portugal on behalf of Scottish Aviation's subsidiary freight and travel agency business, Scottish Air Express Limited, carrying passengers to Lisbon to embark on a fourteen- day cruise with the Anchor Line. Other contracts at this time took Scottish Airlines aircraft as far afield as Canada and the United States, South Africa and the Middle East.

The most hazardous of the airline's early operations was certainly the provision of a service to the Faroe Islands which were equipped with a small single strip wartime runway located on a saddle of ground between two valleys. With hills on either side the only approach to Vaago was straight up the fjord and the difficulties were further compounded by the total absence of radio aids and by the fact that, in stark contrast to Prestwick, the airfield was affected by fog for at least part of 300 out of 365 days in the year. In these circumstances it was scarcely surprising that Scottish Airlines only allowed its most Senior Captains to fly this route.

Yet probably the most dramatic of Scottish Airlines' early ventures lay in the participation of five of its Dakotas along with aircraft from B.O.A.C., Silver City Airlines and Westminster Airways in the airlift and relocation of thousands of refugees during the troubles which followed Indian Independence. During "Operation Pakistan" in September 1947 some 7,000 Hindu refugees were carried from Karachi to Delhi and approximately 1,500 Moslems in the opposite direction. In a similar undertaking, christened "Operation India", which commenced in October a total of 35,000 refugees were carried.

Indains being evacuated from Pakistan crowding into a Scottish Airlines Dakota at Quetta Airport

The experience left the Company well equipped to take part in another airlift which was shortly to take place much closer to home. Scottish Airlines had two Dakotas and three Liberators engaged in the relief of Berlin initially in August 1948 and then for the period 19th February - 12th July 1949. During this time the Company completed 497 sorties, flying 1,514 hours and carrying over three thousand tons of supplies including that most unpopular cargo, petroleum.

Despite the loss of the opportunity to conduct scheduled services in its own right, Scottish Airlines achieved considerable initial success, at least outwardly, as may be seen from its level of activity and rate of growth during its first two years of operations.

In 1946 the Company flew some 633,000 miles, carried over 28,000 passengers and some 69,500 lbs of cargo plus 6,600 lbs of mail. In 1947 these figures increased to 1,480,000 miles, carrying nearly 44,000 passengers, 163,500 lbs of cargo and 36,000 lbs of mail. By the end of that year Scottish Airlines had a fleet of twenty aircraft and employed 85 aircrew personnel. There were, however, some unusual aircraft in the fleet, including an Airspeed Oxford, used for many years for pleasure flights from Prestwick, two Percival Proctors and even a pair of Vickers-Supermarine Walrus Flying Boats, which were contemplated for inaugurating a service between Prestwick and Brodick on the Island of Arran, although they were never actually put to civil use.

One of the unusual aircraft in the fleet, an Airspeed Oxford

As the early charters faded out, the Scottish Aviation Board grew increasingly concerned that its fleet would prove to be too large for its workload. The opportunity which was grasped to avert this situation gave a new dimension to the old idea of operating under a "flag of convenience". Instead of supplementing the capacity of existing operators Scottish Airlines' resources were applied to the creation of two new national airlines in Greece and in Luxembourg.

The prospect of establishing a flag airline for the Greek Government had emerged early in 1947 but progress was impeded by the political complications and it was not until November that an agreement was signed in Athens between Scottish Aviation and the Trustees of the Pension Funds of the Greek Army, Navy, Air Force and Civil Service to form Hellenic Airlines (Hellas) the precursor of Olympic Airlines. The four Pension Funds contributed 60% of the capital while Scottish provided the remaining 40% in the form of four Dakotas and a Liberator. David McIntyre and Tommy Hope, who was appointed Hellenic's Co-Managing Director, were the two Scottish Aviation personnel to become Directors of the new Company, although in line with its shareholding SAL also nominated another four of the remaining thirteen Greek Directors. SAL also provided technical and administrative staff and undertook to recruit Greek employees to be trained at Prestwick in order to take over from the British personnel in due course. In the meantime Scottish Aviation retained completed technical control.

The King and Queen of Greece at the inauguration of Hellenic's Cairo/Athens/Rome/London Service.

Scottish Aviation's involvement in the creation of Luxembourg Airlines was similar although the operation was on a smaller scale. Preliminary discussions took place with the Luxembourg Government in the Summer of 1947 and by the end of the year it had been agreed that the new airline should have a capital of six million francs,

3,500,000 being credited to Scottish Aviation in exchange for a Dakota aircraft while the Luxembourg Government would have a holding of one million francs and the local Directors would contribute the balance. A second aircraft was to be transferred to the airline against debentures. One of Scottish Airlines' Senior Captains, M. R. Lacey, was appointed Managing Director of the new organisation and initially all the aircrews, ground engineers and radio personnel were also British.

Two of the aircraft supplied by Scottish Airlines for Luxembourg

Both Hellenic and Luxembourg Airlines commenced operations in February 1948. The initial Greek services linked Athens with Crete, Larissa, Rhodes and Salonika and provided a non-stop flight to London. By the middle of 1949 these had been expanded both internally within Greece and overseas to link Athens with Alexandria, Nicosia, Paris, Tel Aviv and Glasgow. The original routes from Luxembourg were to Zurich and Paris, but within a fortnight an additional service had been opened to link

Paris and Frankfurt and this was the only air connection between those two cities at that time. In due course this network was also extended, partly through the use of Dragon Rapides supplied by Scottish Airlines to incorporate routes from Luxembourg to Athens, Brussels, London, Nice, Rome and Tel Aviv.
Outwardly therefore these appeared to be extremely successful enterprises but financially the picture was very different. It rapidly became apparent that Luxembourg Airlines had been under-capitalised from the start and that, unless this could be redressed, Scottish Aviation would not see any positive return on its investment. As early as June 1948 the SAL Board was pressing the Luxembourg shareholders to put up more working capital or to liquidate the Company in order to pay off the debt to Scottish.

The relationship with Hellenic was subject to the difficulties of co-management typical of international collaborative ventures, but in this case too there were financial problems. Although the Olympic Games in Athens in 1948 stimulated a great deal of traffic and the internal services were heavily utilised, profitability was poor and again returns to Scottish Aviation on its investment were slow in coming. By March 1949 the outstanding debt amounted to some £60,000 over and above SAL's original capital involvement. The position was further complicated by the unstable political situation in Greece at that time. The Government was forcing a merger of the Greek Airline Companies and while it was granting loans to enable them to repay their debts it was simultaneously controlling foreign currency transactions. As a result the offer of £110,000 which was made to Scottish Aviation early in 1951 to buy out its shareholding was effectively valueless since the Bank of Greece would not then permit the export of Sterling.

The transaction was eventually resolved, but in the meantime the airlines' losses were making Scottish Aviation's overall financial position precarious.

The intrinsic problem with Scottish Airlines was that, in the absence of an opportunity to compete for regular scheduled services, it was dependent on charters which were, even in the case of the associations with Greece and Luxembourg or the successful contracts for the Iraq Petroleum Company, by their nature temporary. Both the Greek and the Luxembourg ventures were undertaken in the knowledge that once Scottish had established the new airlines they would be nationalised.

Forced to pursue opportunities as they arose, the Scottish Airlines fleet found itself either too small - as it was when the prospect of establishing a flag airline in Pakistan coincided with the Hellenic enterprise - or more commonly too big. The heavy standing charges of this large fleet rapidly swallowed up the money which the Company was able to make on its profitable charters, particularly as these were, then as now, frequently seasonal.

The Company's desire to operate scheduled services had never lapsed and the opportunity to do so seemed to re-emerge in 1949 when the British state airline corporations were offering Associate Agreements to independent operators.

British European Airways granted Scottish Airlines an agreement of this sort to fly between Prestwick and the Isle of Man and Scottish subsequently sought permission to operate a service linking Aberdeen, Perth, Edinburgh, Prestwick and Belfast to enable passengers from all these points to make Transatlantic connections from Prestwick, but the final sector to Belfast was not approved and in its absence the service was never inaugurated.

Yet in October 1950 further BEA Associate Agreements enabled Scottish Airlines to operate a number of new schedules. One linked Prestwick with Birmingham, Blackpool, Liverpool and Exeter while another took in Aberdeen, Errol and Edinburgh and others connected Renfrew with Leeds, Newcastle and the Isle of Man and the Isle of Man with Blackpool. The opportunity which Scottish Airlines really wanted was to operate a London schedule and eventually it was agreed with BEA that a Prestwick-Northolt service could be introduced from 9th June 1951. The fact that Scottish Airlines were obliged to make an intermediate stop at Burtonwood to serve Manchester and Liverpool, whereas BEA's own route was direct from Renfrew to Heathrow, simply added zest to David McIntyre's desire to demonstrate that his airline could out-perform the nationalised corporation. He took a delight in the fact that although the two services left their respective starting points virtually simultaneously at 8.15 am, Scottish Airlines, largely by virtue of a fast car service from Northolt, still managed to have its passengers in Central London first, despite their intermediate stop.

It was on this route that Scottish Airlines achieved the distinction of becoming the first British operator to introduce the new safety feature of rearward-facing seats, a development which quickly prompted BEA to approach SAL with a view to having this modification incorporated in its own aircraft.

In-flight refuelling

The service had carried over 27,000 passengers when it was suspended on 14th February 1953. Shortly afterwards the Company sold off the bulk of its fleet of Dakotas. The problem was that the London service was failing to make money and the losses of the airlines as a whole were proving to be a growing drain on the Company's other activities.

The notion of separating the Company's airline activities from its manufacturing business had been contemplated even before the former had been established, but the proposal gained a new momentum as the airline losses began to swallow up the profitability and eventually threaten the very existence of the factory.

In August 1951 the Board called for a detailed plan to enable this separation to be implemented from the end of the Company's financial year on 30th September.

In the event it was agreed to defer the implementation of this decision, which would effectively have meant the end of Scottish Airlines, on the advice of the Company's solicitors. They suggested that the proposed action would serve to prejudice Scottish Aviation's claim for compensation from the Ministry of Civil Aviation in respect of the compulsory acquisition of Prestwick Airport, particularly since the Company's case at that time was founded on the argument that the airport should be valued not simply on the basis of its constituent land and buildings but as a utility and a total going concern.

And so Scottish Airlines survived, but David McIntyre, who had been devoting almost all of his time to the airlines, became increasingly isolated as the champion of this particular branch of the Company's activities. Outwardly at least he remained perpetually optimistic that the next change of Government would produce the revisions in the regulation of British civil aviation necessary to enable Scottish Airlines to operate regular scheduled services and to achieve not only profitability but also the scale of operation of which he always believed it was capable.

Among the ventures which he undertook to stimulate wider public interest in civil aviation and Scotland's role in it was the foundation in 1947 of a subsidiary Company, Prestwick Press Limited, and a short-lived journal entitled "Scottish Aviation" which combined features on aviation, Scotland and general travel and which was modelled on the American National Geographic. He even proposed at one time to exhibit SAL's converted Liberator fuselage in the window of a prestigious Glasgow shop to attract interest in the airline. Both these initiatives were designed to foster a state of "public air-mindedness" comparable to that which he had experienced during his earlier visits to the United States.

Above all he canvassed energetically and endlessly to try to erode the preferential position enjoyed by the State airlines and to secure greater opportunities for the independent operator.

He was never to achieve all that he wished in that respect but he remained a passionate advocate of the Company's involvement in airlines. Perhaps his greatest asset, however, was his ability to charm and persuade others, who might have been expected to take a more harshly pragmatic business view of the situation, into supporting or at least putting up with his attempts to turn his vision into reality.

By the end of 1952 Scottish Aviation's factory business was expanding but the airlines were contracting. During that year Scottish Airlines had carried some 30,000 passengers and flown approximately 588,000 revenue miles, somewhat less than in its first year of operation. Equally the factory was making profits but the airlines were losing money at a rate of £1,000 per week. The situation was considered to be so serious that at the beginning of February 1953 the Board resolved that the airline operations should be terminated as soon as possible and certainly within six months. Within the month, however, the Berlin Airlift had recommenced. McIntyre immediately placed two Dakotas on this work and argued that, with the prospect of £49 per hour of revenue against costs of £32 per hour, circumstances had changed dramatically, profits could be projected and the airline should therefore be kept in operation.

In fact the Company's Bankers, who were pressing hard for the closure of the airlines, had forbidden SAL to take part in the Berlin airlift, but McIntyre's Nelsonian action, like an earlier one by Hamilton and himself on Mount Everest, brought its reward. The Bank was persuaded to relent on its ultimatum and Scottish Airlines was given another lease of life, albeit with stringent conditions, including stipulations that there should be no new capital expenditure on aircraft and no new contracts for airline services without the Bank's express consent. It was also stated that the Airlines were not to continue beyond the end of September 1953.

Ironically, when the time came the airline was making money and so was allowed to remain open, although by this stage the fleet had been greatly reduced.

Liberator's post-war interior

Scottish Airlines supplied BOAC with Liberator crews

The Liberators had all been withdrawn from service by October 1950, thus closing a small chapter of history in itself, for Scottish Airlines was the only private British airline ever to operate the Liberator. Although BOAC operated a thrice-weekly Prestwick-Montreal service using Liberators from 1941 to September 1949, this too was essentially a Scottish Aviation operation. Scottish serviced the aircraft during the war and from 1st April 1949 supplied BOAC with Liberator crews for this route to enable the state-owned corporation to release ten of its own crews for training on more modern aircraft.

Virtually all of Scottish Airlines' thirteen Dakotas had also been disposed of by the end of 1952 when the Company acquired the first of a fleet of eight Avro Yorks which were applied to the trooping contracts which formed the bulk of the airline's work during its final years of operation.

Trooping in

In 1950-51 Scottish Airlines had carried out a number of movements for the Royal Air Force and the Air Ministry, including a monthly service to the Middle East on behalf of the Naafi, but principally moving service personnel, equipment and families. Destinations included West Germany, Benghazi, Ceylon and Gambia. It was on similar work that the Yorks were now engaged. Configured to take fifty passengers, three aircraft entered service in January 1954, based at Stansted, and carrying cadets to Montreal for training in Canada. 1,700 cadets were transported during 1953. After that contract ended in 1954 additional trooping contracts were obtained which took Scottish Airlines to Malta, to Cyprus and to the Suez Canal Zone in Egypt. Because the trooping flights into the Canal Zone were quasi-military operation, the Scottish Airlines crews were obliged to take the unusual step of changing out of airline into service uniforms in the course of the sector from Malta to Suez.

Pre-flight checks?

During 1954 the airline's passenger total, which had held up at around 30,000 during the previous year, had fallen to 24,000 but the revenue miles flown had again increased to over a million.

The trooping contracts were a profitable source of business and it finally appeared that Scottish Airlines might be in a position to operate successfully as a separate entity. From 1st October 1954, therefore, fully ten years after the idea of separating the airline and factory interests had first been conceived, Scottish Airlines (Prestwick) Limited was incorporated with a paid-up share capital of £100. 99% of these shares were held by Scottish Aviation and one remained in the name of David McIntyre who was at this point the sole Director of the new Company.

Scottish Aviation also made a temporary advance of £5,000 repayable in two months' time and transferred the airline assets to the new Company at their written-down value, the parent Company granting Scottish Airlines a loan for this amount which, it was hoped, would be largely repaid during the first year of independent operation.

Despite the chequered financial history of the Airlines in the past, this faith was not misplaced. By the end of the year Scottish Airlines (Prestwick) Limited had repaid £20,000 to the parent Company and had a bank balance of £92,000. The irony was that Scottish Aviation itself was by then heavily committed to investment in the launch of the Twin Pioneer project and its liquidity during the final three months of 1954 had been entirely dependent on the revenues generated by Scottish Airlines.

The inherent uncertainty of the charter business was, however, never far away and 1955's operating profit of £32,000 was succeeded by a loss of £41,000 in 1956. The airline's problems were compounded by a series of accidents to the Yorks. Strangely the only accidents to the Scottish Airlines fleet during its history were on this type. Between 1954 and 1957 five of these aircraft were written off in crashes, three at Stansted and two at Luqa on Malta, and the York was eventually prohibited from carrying passengers.

But it was the crash of Scottish Aviation's Twin Pioneer demonstrator in the Libyan desert in December 1957 which marked the beginning of the end of Scottish Airlines, for with the death of David McIntyre the unique driving force behind this particular operation was lost. The writing was on the wall when the trooping and Ministry freight contracts came to an end and by the end of 1958 only one Dakota, G-AMPP, was left in service.

This was employed chiefly in 36-seat configuration on a seasonal scheduled operation between Prestwick and the Isle of Man, a service which the Company had provided since 1948 and which remained busy, sometimes involving as many as three round trips each day. This service finally closed for the last time on 25th September 1960.

Appropriately, however, Scottish Airlines' operations ended where they had begun, on the Prestwick-Belfast route. When "Papa-Papa" landed at Prestwick from Northern Ireland at 10.33 pm on 9th November 1960 it had completed the final commercial flight flown by Scottish Airlines.

Just as the remaining Yorks had been offered for sale to Skyways in 1958, so Dan-Air Services at Gatwick now bought Scottish Airlines' last Dakota and with it the scheduled service from Prestwick to the Isle of Man.

Air
Charter

The Scottish Airlines charter fleet is immediately available for journeys within the United Kingdom or overseas—for business travel or for a quick trip to a race meeting, a football match, or a re-union.

Note the telephone number
Prestwick 7281

SCOTTISH
AIRLINES

Prestwick Airport

1947 advertising

Scottish Airlines had not achieved all that its founding spirit had wished but in this, as in other respects, David McIntyre's aviation vision had run ahead of the political and financial pressures to which it would be subject. Some of the work on which the airline found itself engaged perhaps lacked the grandeur of the original vision but it never wanted for variety. Charters ranged from the exotic to the mundane to the frankly eccentric, including as they did the movement of Indian dancers and their stage properties from Paris, a long-standing milk run from Northern Ireland to Liverpool and even the transportation of goldfish from Northern Italy.

Yet Scottish Airlines operated for fifteen years. In that time it contributed directly to the foundation and operation of scheduled airline services, not only within Great Britain but in several other European countries. It transported thousands of servicemen and their families around the globe. It took part in the vital relief operations in Berlin and India and it provided a service to a multitude of charter customers. While the scale of its operations may never have fully reached its founder's aspirations, the nature of its activities undoubtedly did. "The World O'er" had been turned from a statement of ambition into a statement of achievement.

Scottish Aviation Dakota conversion, corporate interior.

Chapter 5

"CONVERTING" The Transition to Peace

War and the prospect of war had stimulated the rapid development of the Company's manufacturing activities during its early years, but by the same token converting to the circumstances of peace was inevitably a difficult process.

It was entirely consistent with the Company's other ambitions that it should wish to design and manufacture its own aircraft. In addition to his plans for a Scottish Aviation "Concord", David McIntyre believed that there was a neglected requirement for a 20-30 seat aircraft with a range of 1200-1500 miles and in December 1944 the Company submitted a proposal to the Ministry of Aircraft Production to design and manufacture a prototype.

The Ministry showed no enthusiasm for this particular initiative but it became evident that Scottish Aviation had established its credentials as a manufacturer when in February 1945 it was asked to submit a design study for a small military communications aircraft in response to Air Ministry Specification A4/45. The Company was given a mere sixteen days to submit its study and was jubilant when the Scottish Aviation design was eventually preferred over the submissions made by Miles, Folland, Heston and Cunliffe Owen. Unfortunately the Government did not then pursue the matter with anything like the same urgency which it had previously required of the prospective contractors.

Nevertheless in July 1946 McIntyre was still optimistic that the order would be forthcoming and that it might be for "anything up to 400 aircraft". The scale of these expectations shows the importance of the Light Communications Aircraft in the Company's post-war manufacturing plans, but the recession in the Industry after the war was to be even more severe than the Company had anticipated. At the end of August 1944 the British Aircraft Industry had employed a total of 1,700,000 people and had outstanding orders for 43,000 aircraft. One year later these numbers had declined to 993,000 employed on orders for 10,300 aircraft and this process of contraction was to continue.

With the rapid cancellation of its Ministry of Aircraft Production contracts, employment at Scottish Aviation fell from a peak of nearly 6,500 in all its activities in 1944 through 3,300 by the end of 1945 to a total of only 1,750 in 1948 when the Air Estimates for 1948/49 were cut from the previous year's level of £214 million to a new figure of £173 million. Consequently the option on the Light Communications Aircraft was not taken up.

Salvation lay in the expertise which the Company had acquired on American aircraft during the war. In the absence of the opportunity to manufacture an entire aircraft to its own design Scottish turned to designing and producing conversions, another sphere in which it had accumulated an immense amount of experience. In fact when the war ended, the Company was engaged on the conversion of 45 Liberators for Coastal Command at a rate of four per month and had a total of 140 to be converted for Transport Command. With a prospect that this work might continue for as much as eighteen months beyond the end of the war there was for a time a real worry that this programme would leave no factory capacity available to undertake commercial contracts.

However, the Ministry Contracts fell away sharply in 1946 and the way was clear for civil conversion work. In anticipation of this Scottish had taken steps to consolidate its war-time role as the design authority in Great Britain for Dakotas and Skymasters by seeking appointment as Douglas' agents after the war.

Dakota conversion, a bridge from war to peace

In August 1945 a Licence Agreement was concluded after which Scottish Aviation was able to advertise itself as the Douglas Aircraft Corporation's authorised repair, overhaul and conversion centre for Great Britain and the first specialist centre of this sort in Europe.
The Company's experience on Dakotas and other American types must have spoken for itself but it would also seem that Donald Douglas, himself of Scots descent, felt a close personal affinity with SAL and with its founder-directors. When the Duke of Hamilton and David McIntyre arrived at Los Angeles for discussions with the American corporation, Donald Douglas' personal pipe band was there to greet them off their aircraft. Unfortunately on landing, the aircraft overshot the point at which the bandsmen were assembled and they promptly set off at the double down the runway pursuing the aircraft which ended up with its nose in the ground and with its two distinguished passengers having to clamber out as best they could. It was a memorable start to a long, memorable and subsequently smoother association.
Yet for a while it seemed that of the various aircraft on which the Company had regularly worked during the war the one for which there might now be most civil interest was the Catalina flying boat. This seems curious today until one recalls the extent to which flying boats had dominated Britain's pre-war civil air routes. Scottish Airlines tried to acquire a number of Catalinas for their own operations and inquiries were also received for similar conversions from Denmark and the United States and, at least in the case of one aircraft for Icelandic Airways, the conversion was actually carried out. The success of the transatlantic air ferry of landplanes in

The authorised repair, overhaul and conversion centre for Great Britain.

which Scottish Aviation and Prestwick had played such a large part meant, however, that the flying boat's heyday was now over.

Although thwarted in its attempts to build 'Concord' utilising its re-design of the Liberator fuselage, SAL did offer a conversion of the Liberator LB30 into a deluxe 24 seater civil airliner, and this configuration was used by Scottish Airlines.

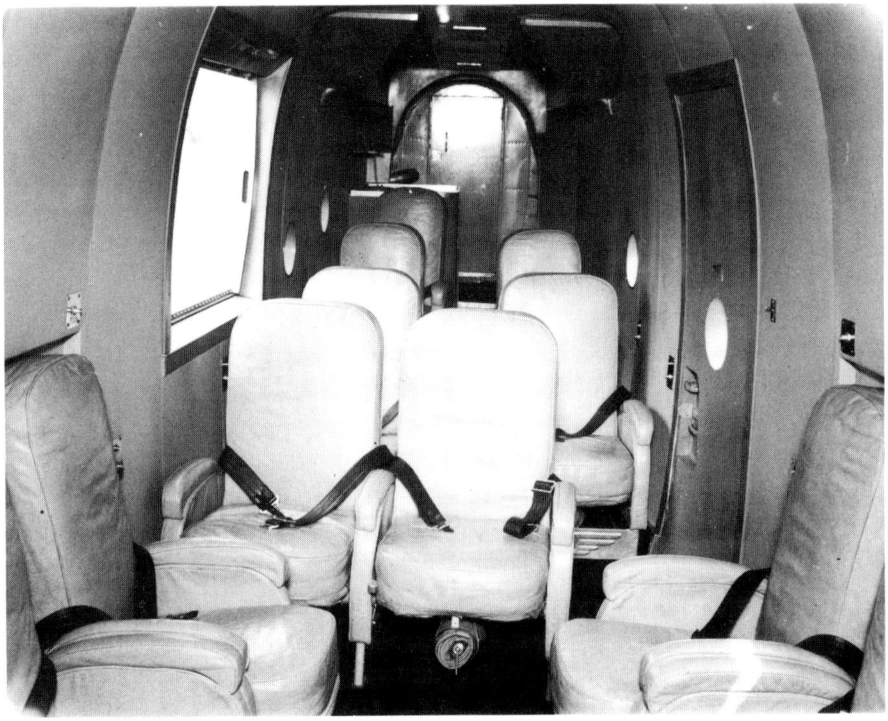

Interior of an early Liberator conversion, 1946

Under this arrangement the aircraft was substantially re-worked to replace bomb doors, catwalk and structural members with a new under-fuselage providing a continuous passenger cabin with no change in floor level. Both the nose and tail sections were streamlined to improve performance. The plexiglass nose was replaced by a loading door into a forward freight compartment, while the tail section was modified to accommodate passengers' baggage. A full size passenger door was provided on the port side of the fuselage while the beam gun aperture on the starboard side was replaced with a plexi-glass panel to provide an unusually large observation window. In addition to soundproofing, buffet and toilet facilities complete with the modern convenience of power supply for electric razors, passengers were provided with individual lighting, cold air louvres and emergency oxygen supply.

The Company also offered a C54 conversion as a deluxe 44-seater, but it was the Dakota which was to be the mainstay of its conversion programmes. These former military C-47's were sold in a choice of standard or deluxe 21-seat airliner or alternatively as a special 'corporate' conversion for "Business Executives and Private Owners".

Principal structural alterations in the deluxe airliner model included the deletion of the navigator's compartment to provide luggage space, the replacement of the metal cabin floor with reinforced balsa wood, the re-positioning of the toilet and the provision of a new bulkhead to form an aft baggage compartment with a small loading door on the starboard side. In addition the existing double freight doors were removed and replaced by the small passenger door which was standard on the civil Dakota, the DC-3. Emergency measures were supplemented by the installation of fire extinguishers and first-aid kit plus the embodiment of a third escape hatch. SAL also offered to provide life-jackets at an extra cost of sixty shillings apiece.

The interior of the cabin was sound-proofed and the upper portion covered with doped aeroplane fabric, while the lower half from the curtained windows to the carpetted floor was finished with plywood panels upholstered in leather of the customer's chosen colour. Magazine containers, hand-rails, luggage panels, adjustable seats with integral ash-trays, call buttons and cold air louvres were all incorporated. Gun port windows were replaced with perspex. A buffet was fitted at the rear of the passenger compartment with an optional sink unit. As a final customer option the Company also installed racks for thermos flasks. Customers were asked to specify the dimensions of their particular flasks.

For its executive models the Company undertook to arrange the passenger cabin according to the individual customer's specification, dividing the space as required into sitting room and bedroom with the facility to transform the latter for daytime use either as a lounge or as a buffet-bar complete with cocktail cabinet, icebox and cut crystal decanters. Desk, filing cabinets and standard office furnishings could also be installed, as could full facilities for a galley incorporating refrigerator, electric stove, water filter, pantry and sink. A second toilet and fitted wardrobes were additional customer options.

The main source of supply for surplus Dakotas was the United States War Assets Administration and the principal storage unit for these aircraft was at Silloth airfield on the Solway Firth, only thirty minutes flying time from Prestwick and a site which Scottish itself had used as a dispersed repair unit for Liberators and Dakotas during 1944/45. The brochure advertising Scottish Aviation's conversions drew attention to the Company's ability to provide clients with prompt, first-hand advice on the most suitable aircraft available for purchase. Among the services which the Company offered to potential customers were the provision of reports by its Service Engineers on the general condition and running times on the engines and airframes of the aircraft for sale, arrangement of the purchases through the Foreign Liquidation Commission in London and even collection of the aircraft from Silloth. These services were provided free of charge provided that an order for conversion or overhaul was received.

Deliveries were quoted on the basis of six to eight weeks from receipt of aircraft at Prestwick in the case of the standard Dakota conversion, ten weeks for the deluxe and ten to twelve for the executive model. For the C54 this increased to between twelve and fourteen weeks and in each case customers were advised that in the event of a major overhaul being required, they could expect the cycle to take an additional two or three weeks, depending on the condition of the aircraft.

Customers for the Company's conversions were not slow in coming forward. When David McIntyre visited Holland in November 1945, KLM expressed its interest and even made a tentative suggestion that Scottish Aviation and KLM might establish a combined manufacturing unit for large conversion work. An order for five aircraft for the Royal Dutch Airlines was the first of many conversion contracts won by SAL. The list of subsequent customers confirmed the accuracy of the Company's prediction that airlines would be dependent in the first instance on American war surplus aircraft for their post-war equipment. By any standards the length of the list is impressive. Airlines for whom aircraft were converted at Prestwick by Scottish Aviation included KLM, BEA, Swissair, Aer Lingus, Transcontinental and Western Airways, American Overseas Airways, TAP (Portuguese Airlines), Sabena, Norwegian and Danish Airlines (later to be amalgamated as Scandinavian Airline Services), Ceylon Airways, Cyprus Airways, Abyssinian Airlines, Bharat Airways, Orient Airways of Pakistan, Olympic Airways, Hellenic Airlines, Luxembourg Airlines, Icelandic Airlines, Czechoslovakian Airlines, Algerian

Airways, Syrian Airways and many charter operators including Channel Airlines TI, Jersey Airways, Westminster Airlines, Global Aviation (which was owned by former American Ferry Pilots), Fred Olsen Lines of Norway, Blue Sky in France and Alpar SA of Switzerland.

Customers

The growth of this list was a measure of the reputation which Scottish Aviation's work had acquired. Perhaps the best indicator of this reputation, however, was given when a Portuguese gentleman presented himself without warning at the Main Gate of the Prestwick Factory and asked to see Mr Neill, the Assistant General Manager. He was duly shown in and within the hour had not only signed a contract for the conversion of five Dakotas at a price in excess of £150,000 but had tabled an envelope containing Escudos to the value of £20,000 by way of a deposit.

There were also some notable sales of the executive model. Lord Beaverbrook, who knew Scottish Aviation and its capabilities well from the war, asked the Company to convert a Dakota into a VIP flying stateroom for him. The result was widely admired and Churchill was a frequent passenger in the aircraft. The resulting publicity prompted orders from other VIPs, notably the King of the Belgians, the Maharajah of Baroda, the Prime Minister of India, who requested what was believed to be the first pressurised Dakota, and Haille Selassie, Emperor of Abyssinia, who was so specific in his requirements that he went so far as to send his own personal painter over to Prestwick to paint the Lion of Judah insignia on his aircraft.

The Emperor's personal painter

The association with Douglas therefore proved to be a most important bridge into the post-war period for the Company's manufacturing business. Pursuing related initiatives Scottish Aviation also became in December 1945 distributor and official repairer of Pratt and Whitney engines and early in 1947 was negotiating to make a similar arrangement with the Sperry Gyroscope Company, which was withdrawing from Prestwick the representatives which it had stationed at the airport during the war.

Difficulties in the supply of test equipment delayed the establishment of the engine shop, but the value of this new venture was readily demonstrated when the Company found that with a relatively minor overhaul it could restore war surplus engines to a condition where they would sell for as much as ten times the price which they had cost to acquire.

The profits which could be achieved on the Dakota conversions were not on the same scale but the converted aircraft could be sold for nearly three times its original cost to the Company and this contributed very largely to the Company's survival during the difficult years after the war.

In July 1946 David McIntyre had envisaged these conversions as essentially a stop-gap activity until such time as the Ministry confirmed a substantial requirement for the Light Communications Aircraft, but in the event the Dakota conversions were to continue off and on for many years, reflecting the in-built quality of the design of this outstanding workhorse aircraft, still to be seen occasionally flying at Prestwick during 1985.

The Dakota was a great friend to Scottish Aviation over the years, although there came a point when this friendship was nearly over-stretched when the aircraft came close to usurping the name which McIntyre proposed to give to the civil variant of SAL's Light Communications Aircraft. This was in 1950 when the factory undertook the conversion of 38 BEA Dakotas to a 32-passenger standard. BEA wished to call the modified aircraft Pioneers, but at McIntyre's insistence they were eventually christened "Pionairs".

The pace of the Dakota conversion activity accelerated rapidly and in March 1947 the Company seized the opportunity to make an agreement with the Foreign Liquidation Commission to take approximately 40 of the aircraft which could not be accommodated at Silloth and hold them at Prestwick. The Company's responsibilities were to inhibit the engines and to look after the aircraft, but with no liability for loss, damage or deterioration. In exchange for what was effectively a parking agreement, the American authorities agreed that as soon as the first aircraft had been sold to an outside customer, then Scottish could have three for itself free of charge. Finding customers was not a problem. By November the Company's difficulty was that there simply were not enough aircraft to satisfy demand.

The "Pionair" conversion

The resilience of the Dakota conversion business was remarkable, but when this work began to pass its peak the gap in Scottish Aviation's manufacturing workload was acutely felt.

In seeking to fill this gap the Company's preference was to remain close to its original reasons for foundation and to find other aircraft work, but at the same time the urge to remain as far as possible master of its destiny was still strong. On the other hand the investment required to undertake a new aircraft project, although modest by today's standards, was still formidable and Scottish was unable therefore to pursue the invitation which it received from Blackburn Aircraft in March 1946 to take part in the manufacture of a 320,000lb flying boat, even though this was similar to a design concept on which it had already done a study of its own.

One association which the Company was able to pursue, however, was with the Bell Helicopter Company with which, in 1947, SAL formed an organisation called the Irvin Bell Helicopter Sales Company Limited. 75% of the new Company's capital was put forward by Scottish Aviation and the remainder by the Irving Air Chute Company and Bell. Its objective was to demonstrate and sell the latest Bell Helicopters in exchange for which Scottish would receive a 15% commission on sales and the revenues from crew training for which it would be responsible. Noel Capper, SAL's Chief Test Pilot, and two Scottish Engineers consequently found themselves in Buffalo on an eight-week training course.

The Bell Helicopter, 1947

Filled with enthusiasm for the helicopter when it eventually arrived back at Prestwick with its newly-trained pilot, David McIntyre suggested that Capper should demonstrate its capabilities by landing it on the flat roof of the main administration block in the Palace. Taking the sensible precaution of researching the proposed landing ground beforehand, Capper put his foot straight through the ceiling of the Works Manager's Office and declined to follow suit with his helicopter.

The helicopter business was disappointing. The product was expensive in relation to fixed wing aircraft especially with a busy market in war surplus equipment and although the helicopter's crop-dusting capability aroused some interest and resulted in demonstrations over the sugar plantations of Trinidad and the beetroot fields of France, sales were few and far between.

In the meantime the fluctuating factory workload coupled with the heavy demands placed on the Company by the equipping and operation of its own airline activities were causing an increasingly worrying financial position. Early in 1947 the Board of Directors was expanded by the addition of Sir Ernest Lemon, whose production background and extensive contacts would, it was hoped, strengthen the Company's manufacturing business.

As the Company's liquidity crisis deepened it found itself forced to turn away from aircraft work because of the unacceptably long term and high development costs involved. Instead it looked for projects which would entail relatively little investment and offer a quick return, thereby enabling it to weather the financial storm and to keep the factory's capacity employed and its skill base preserved for aircraft work in the future.

In pursuit of this policy Scottish Aviation again embarked on a series of ad hoc projects ranging from the manufacture of "mushroom" folding caravans which expanded by means of a central screw jack from a 4' x 6' trailer to a caravan which could sleep six people, to the production of aluminium step ladders and even a batch of 10,000 fish boxes. But although some of the ventures which were explored had the merit of offering a prospect of sustained, high volume production, such as the manufacture of Ferguson tractor mower units, a contract which was expected to last for two-and-a-half years and to achieve an eventual production of 44 units per day, the capital investment in machine tools which would have been required to undertake these projects made them, in the circumstances in which the Company then found itself, prohibitively expensive.

The most successful of the interim projects was also of an agricultural nature, and was the inspiration of Dick Ellison, one of the original Flying Instructors at 12 EFTS and by this time General Manager, who was himself a small-holding farmer. This was the project to produce Scottish Aviation-designed tractor cabs, fabricated from aluminium alloy so that they would be sufficiently light-weight to be installed by the driver alone. As the Company's promotional brochure explained,

> "It is appreciated that many farmers, who realise the advantages of tractor cabs under difficult weather conditions, prefer to drive a standard tractor without cab during the better weather months, consequently particular attention has been paid to making this cab easily removed. Spring clips require to be undone and the complete cab can be swung back and off in one minute. Remounting can be done in a similar fashion and in the same time".

The Scottish Aviation light-alloy tractor cab

.... the finished product in place

The display of the Scottish Aviation tractor cab at the Ayr Show in April 1948 received an encouraging reception and suggested that there might be sufficient demand for production to be at a rate of fifty per day. The cab was extremely successful. It was manufactured for many diferent types of tractor and eventually it was even sold in a do-it-yourself kit form.

By the time the Company had exhibited the cab at the Royal Show in July it had added roof trusses and farm gates to its range of agricultural products and had gone so far as to consider making its own combined harvester.

The new product in which the Company resided its greatest hopes at this time was not, however, a piece of agricultural equipment but it was associated with transport. It was the light alloy bus body.

The idea was to adapt the Company's aircraft design experience to the building of bus bodies of the stressed skin monocoque construction which had by then become standard practice in aircraft engineering. Light alloy was used throughout, except for items such as chassis mountings and bolts, in order to take advantage of its greater resistance to corrosion and so produce longer life and fewer maintenance problems. In addition, of course, the high strength characteristics of the light alloys made it possible to make the vehicles as strong as their traditional steel and timber counterparts but with a substantial reduction in weight and consequently a considerable saving in fuel and therefore operating costs.

So it was that the Prestwick Stress Office found itself calculating load factors not on aircraft designs nor on airframe conversions but on a series of bus bodies from the luxury coach to the standard double-decker, the low bridge variant and even an "overseas" model, ideally suited for tropical climates because, as the promotional brochure pointed out,

> "The light alloy structure is, with the minimum of protective treatment, completely resistant to corrosion and impervious to insect attack. The various fabrics used can be 'tropicalised' and rendered rot-proof where humidity of climate makes such precautions necessary".

Again considerable interest was shown in the Company's new product, by SMT, Albion Motors and the Glasgow Corporation amongst others. Initially sales were promising and the Company's particular expertise in working with aluminiium alloys on aircraft meant a profitable source of repair work for which the customers themselves were not equipped. By April 1948 Scottish Aviation was planning on achieving a rate of production of two bus bodies per week by the end of May and of rising to a mature rate of ten per week, representing an annual turnover of £1,000,000 in itself. These ambitions were, however, to be frustrated. This was partially as a result of legislation which first restricted the allocation of bus chassis and then, with the introduction of a new petrol tax in April 1950, severely depressed the home market for coaches. To some extent also the eventual failure of the bus body programme sprang from the sheer cost of the aircraft-type materials which were being employed. Despite the advantages of fuel economy, easier maintenance and above all durability, which was borne out by exceptional lengths of service, the bus companies were not prepared to pay more than they would have had to pay for a traditional steel and timber body.

Light Alloy bus body construction

Customers included Glasgow Corporation

Nor was the profitability of the project helped by the sheer range of bodies which the Company put on offer. In attempting to secure the high volumes of work which it needed to keep the Prestwick factory occupied, Scottish undertook to make models available for AEC, Albion, Commer, Austin, Bedford and Foden chassis, all of which added to the costs of the project.

Scottish Aviation bus bodies

Consequently neither the scale nor the returns of the bus body programme ever achieved the levels which were required to resolve the problem of the deteriorating financial position with which the Company was faced by the late 1940's.

A complete transport package

"Grounded" : bus body production in the Palace

As the Company celebrated its thirteenth birthday in the Summer of 1948 it seemed as if its luck was about to run out. Its bank overdraft was at its limit, creditors were overdue and it was a struggle to find sufficient funds to meet each week's wages bill. It looked as if the Company would be unable to meet its total financial commitments for any longer than about a month.

The Ministry of Aviation, Members of Parliament and the Secretary of State were all lobbied in an attempt to secure work, and Prestwick's strategic location and previous track record were both acknowledged. In the midst of the Cold War which had occasioned the blockade of Berlin there was much talk of future military aircraft work but the immediate position remained uncertain. Furthermore, in the absence of adequate working capital Scottish again found itself in the position of having to turn down the opportunity of participating in a major aircraft project, in this case a collaboration with the Cunliffe-Owen Company to manufacture 50 of its 10/12 seat 'Concordia' feeder aircraft. Approaches made directly to De Havilland and English Electric were more productive, since these had been accurately identified as the two UK aircraft manufacturers with the most work on hand and consequently the best prospects for suitable sub-contracts. In due course these approaches produced a contract for the refurbishment of 45 Meteors. But in the meantime there was too little work coming forward too slowly and the financial situation reached crisis point.

The Trustees of the Hamilton and Kinneil Estates had become increasingly apprehensive about the position. In May 1947 their failing confidence in the Company's prospects prompted them to seek to recall their vital £50,000 loan, although they were subsequently persuaded to extend the loan pending a settlement in respect of the Government's compulsory acquisition of Prestwick Airport. It was an uneasy truce. With no visible improvement in the position twelve months later the Hamilton Trustees called for a detailed analysis of the Company's indebtedness from an independent party and indicated that this might be expected to confirm their view that the Company should go into voluntary liquidation. By September they were asking the Company bluntly whether or not the Ministry of Supply would support it with work. If the answer was in the negative, then the Trustees proposed to appoint a Controller of Scottish Aviation Limited "for the protection and gradual realisation of that Company's assets".

These desperate circumstances prompted consideration of desperate remedies. Thought had already been given to declaring a number of the large hangars in the Bl Scheme redundant, although David McIntyre had strongly resisted this, pointing out the danger of another aircraft company installing itself and taking advantage of Prestwick's unique assets. In May 1950, however, the closure of the "Palace" itself was even contemplated.

An alternative solution had offered itself briefly in July 1949 when the Douglas Aircraft Corporation had asked SAL to undertake conversions of the Super DC-3 and the possibility of Douglas acquiring a shareholding in Scottish had been discussed.

But in the event it became clear that the proposed conversion would be too expensive to generate a large market and the prospective union with the Douglas Aircraft Corporation fell through.

With the Company which he had worked so hard to develop finding itself under threat, David McIntyre's response was characteristic. He proposed to the Hamilton and Kinneil Trustees that as an alternative to liquidation they should either allow him the opportunity to raise the finance to repay their loan and relieve them of their shareholding or at least permit him to continue a portion of the Company's activities relative to the scale of his own shareholding.

There are signs that the old friendship between McIntyre and Hamilton came under considerable strain at this time, but it was clearly the Hamilton Trustees rather than Douglas-Hamilton himself who were pressing for liquidation and there seems to be no doubt that it was the Duke who repeatedly persuaded the Trustees to stay their hand. They did so, not simply for a matter of months, but over several years. During one remarkable period of ten months between September 1948 and July 1949 no Board meetings took place.

McIntyre was never able to raise the capital necessary to go it alone, but the breathing space which he was granted as Managing Director gave him the opportunity to continue to apply his enormous energies to an issue which he had pursued with singular determination since the beginning of 1946, the achievement of a satisfactory settlement of the Company's claim for compensation for the loss of the Airport.

In one sense the compulsory acquisition of the Airport which had been declared by the Ministry of Civil Aviation with effect from 1st April 1946 was simply an extension of the existing situation into the post-war period. In another respect it was the completion of the process by which external forces usurped the Company's control of the airport. Even under requisition many of the airport services at war-time Prestwick had been provided by SAL but there was no counterpart to this earlier station contract after the Ministry of Civil Aviation declared itself to be the new landlord.

But although the Government had requisitioned the airfield in 1941, the requisition rent had never been agreed. After the war the Ministry offered to lease the factory property to the Company but Scottish Aviation was not prepared to agree or accept these terms pending payment of compensation for its loss of ownership.

This settlement was still outstanding at the end of June 1947 when SAL demanded that the Ministry of Civil Aviation should immediately either de-requisition the Company's property or acquire the Airport holding and arrive at a tenancy agreement or introduce a temporary agreement to regularise the position pending negotiations. In the uncertainty of the existing situation the Company could not even authorise major building repairs and the condition of some of the factory's roofs had reached the stage where the buildings were on the verge of becoming uninhabitable.

At this point the Company was prepared to settle the issue out of court and, although the Directors believed that the figure was substantially below the full value of the asset, to accept a minimum compensation of £300,000, provided this was paid immediately.

But this offer was not forthcoming. The figure which the Ministry proposed was a mere £87,000. The Company's view was that the Airport should be valued not simply in terms of its constituent land and buildings but as a going concern. Nevertheless, the pressures of its financial position were such that in March 1948, with the Hamilton and Kinneil Estates pressing for the repayment of their loan, the Duke of Hamilton and Sir Ernest Lemon arranged to meet the Permanent Secretary at the Ministry of Civil Aviation to seek an immediate payment to account of £250,000 with a minimum of £200,000. There is little doubt that, if the Government had offered this sum in full and final settlement at this point in time, the Company would almost certainly have accepted it.

But again the Government was not prepared to come to terms and it was not until the end of May 1948 that it revised its offer to a new figure of £175,000 although this included £44,000 for buildings erected under the Air Ministry's Sinking Fund Scheme which had still to be confirmed as being owned by the Company. The question of SAL paying the residual value of the buildings also remained to be addressed. In the absence of a satisfactory settlement the Company took the matter to litigation in 1950 and the peculiarities of Scottish law introduced further complications into the story. On the advice on its solicitors the Company argued that because the Civil Aviation Act of 1946 had been repealed and because, in the drafting of the 1949 Act, Scotland had been omitted from the section giving the Minister power to acquire property, the Ministry of Civil Aviation could not compulsorily purchase Prestwick Airport. It was hoped that this embarrassing situation would prompt the Government to resconsider an out-of-court settlement.

By March 1951 the Claim which the Company had lodged contemplated three possible results and set a value against each. First, for Land and Buildings and closure of the Company's business - £1,983,675. Secondly, for Land and Building and reinstatement at Prestwick - £798,080. Finally, for Land and Buildings and reinstatement at Heathrow - £1,008,780. In each case the land was valued at £223,080 and the buildings at £258,000, a total of over £481,000.

In May 1951 the Ministry confirmed an offer of £176,000 in full satisfaction of the claim for ground and buildings but by this stage the Company's manufacturing business was starting to pick up and, although the liquidity position remained precarious, the Directors decided to reject the offer as "utterly unacceptable".

By November 1951, when the Earl of Selkirk, one of SAL's original Directors, resigned to take up a Government appointment, the Company was feeling sufficiently confident to press the Ministry to choose one of three alternative outcomes. These were either a negotiated settlement for the purchase of the land and buildings combined with a 99-year lease of the factory or this same arrangment plus a management contract for the operation of the Airport in line with State requirements or, thirdly, for Scottish Aviation to retain the ownership of the land and buildings and to operate the Airport under a Management Contract.

In the end it was not until November 1953, after the valuation of the land and buildings had been referred to arbitration, that negotiations with the Ministry of Civil Aviation were brought to a conclusion.

The struggle had taken nearly eight years. But the outcome was a great personal triumph for David McIntyre. The final settlement provided that all claims by the Ministry against Scottish Aviation should be discharged, as should all the reciprocal claims, leaving a balance of some £6000 to be paid by the Company to the Ministry. This meant that in addition to the Arbiter's award of

£343,835 the Company had secured the value of the outstanding rent and other charges owed to the Ministry, which amounted to some £100,000. In effect the Company had achieved a settlement of £450,000 for their total claim together with a 99 year lease.

There had been some anxious moments in the meantime and never more so, perhaps, than in 1951. In January the Board had agreed in principle that unless the Company could arrest its losses and start to move back towards profit that year then it could not continue.

In those days the Directors held their Meetings not only at Prestwick, but in a variety of other locations including Glasgow, London and frequently Edinburgh, where a common venue was the Palace of Holyrood House of which the Duke of Hamilton was the Hereditary Keeper. In view of the Company's uncertain financial position at that time it must have seemed disturbingly apt that, of the many famous paintings in the building, the one which dominated the room in which the Board meetings took place was "The Massacre of the Innocents".

Among those who were not prepared to see the Company massacred without putting up a fight was T.D.M. Robertson who had been recruited in 1951 as part of Scottish Aviation's continuing efforts to strengthen its management team. He came from Hawker Aircraft at Kingston where he had been Director and General Manager to fulfil a similar role at Prestwick as a Director and as General Manager of SAL's Design, Administration and Manufacturing activities.

He had very clear views on the strategy which the Company needed to pursue in order to survive. First it had to get rid of the unprofitable, ad hoc contracts with which it had encumbered itself, however large and superficially attractive these were. Then it had to restore its skills and re-position itself in the mainstream of the aircraft industry even if this were as a sub-contractor in the first instance.

The plan was pursued with a boldness which, while it concealed SAL's financial weakness from the outside world, reflected an understanding within the Company that it could not afford to fail. British European Airways, one of the Company's largest customers, were advised that unless they were prepared to re-negotiate the terms of their conversion contract Scottish Aviation was not interested in doing any more of this work for them. What was more it quickly became apparent that the Company meant what it said. The terms were re-negotiated, satisfactory margins were achieved for the first time and most importantly, perhaps, the whole process was carried out in such a way that BEA went on to award further conversion contracts to SAL on the Viking.

A number of his colleagues found some of the new General Manager's actions hard to swallow. The old Fokker airliner, which had served during the war as the flying classroom and subsequently in Scottish Airlines livery, had come to rest in the hangar which bears its name to this day. Although at the end of its life, it remained one of David McIntyre's treasures. "TDM" had it removed from the hangar and parked outside on the tarmac to make way for the new BEA conversion contract. McIntyre was at first furious at this heresy but eventually came to terms with the fact that in this case at least it was necessary for the old to give way to the new. An attempt was even made to build him a cocktail cabinet out of the aircraft as a memento, but in the event it transpired that most of the wooden structural members were in fact laminated and quite unworkable.

Other victims of the General Manager's critical review of the Company's manufacturing programmes were the Albion Bus Bodies and a £100,000 machining contract for the United Kingdom Atomic Energy Authority. When "T.D.M." cancelled this massive contract on the grounds that it could never be made to do anything for Scottish Aviation other than lose money, the Earl of Selkirk was prompted to ask, "Are you sure you know what you're doing?"

The factory's results were the best answer to the question. By September 1952 it was again employing 500 craftsmen and direct production operators and its annual sales turnover had been increased to £800,000, double the level of three years earlier. 1951's £22,000 operating loss had been turned into a £20,000 profit and this was to be trebled in the following year.

In place of the unprofitable contracts which had been weeded out came new work, including an order for 150 pairs of Canberra rudders from AV Roe, Vampire components for De Havilland and engines for Rolls Royce, initially Merlins, Goblins and Gypsies in an association which was projected to last for five years and eventually lasted for over thirty.

As always the Company's changing fortunes also owed a certain amount to chance and, in one instance at least, to the celebrated Scottish hospitality. This was when a Ministry negotiator, who had come to Prestwick to discuss the Light Communications Aircraft, arrived hungry and bleary-eyed having been without sleep or food during a particularly difficult journey from London. His Scottish Aviation host promptly arranged for him to have a large brandy and a room in the Airport Hotel and re-scheduled their meeting to take place over lunch in the world famous Airport Restaurant.

By way of gratitude the Ministry man suggested that at a forthcoming meeting in London on the Light Communications Aircraft, Scottish should take the opportunity to volunteer its services to undertake sister design work on a series of American aircraft which were to come into British service. The result was an extension of the work which SAL had already carried out on Washingtons to include major conversion programmes for Skyraiders, Corsairs and one hundred Avengers.

By May 1954 the Hamilton and Kinneil Estates loan had been repaid in full. Surplus aircraft from the airlines business had been sold off and the factory was again making profits. The Airport acquisition claim had been settled. Overall the Company had moved from being £250,000 in the red to being £625,000 in the black.

Once again the means were available "to continue the initial policy of the Company towards establishing employment opportunity in Scotland".

Precisely where the Company was to go in pursuit of these objectives sprang from one of those fascinating coincidences from which history is made.

Revived, the Light Communications Aircraft

The scrutiny of the Company's various outstanding contracts had revealed that under the terms of the original contract for the Light Communications Aircraft the Ministry of Supply technically still owed Scottish Aviation money. Simply in an effort to obtain this cash the Company pressed for payment. The Ministry acknowledged its liability but demanded for its part that Scottish should fulfil its side of the contract and should proceed to complete the four aircraft which it had been engaged to produce.

Meanwhile the second half of the coincidence which was to determine the Company's immediate future was developing on the other side of the world in the jungles of Malaya where the threat of Communist infiltration was dramatically to resurrect the requirement for a light military communications aircraft which could keep in touch with isolated troop contingents.

Suddenly Scottish Aviation was faced with the opportunity to pursue its preferred path of self-dependence, of manufacturing entire aircraft of its own design and thereby "lifting the Company out of the category of a fringe firm and establishing it as a recognised member of the British Aircraft Industry".

The designs with which it planned to achieve this were names which were to become and remain synonymous with Scottish Aviation: the Prestwick Pioneer and the Twin Pioneer.

This painting hung for many years in the Prestwick factory's entrance hall.

Chapter 6

"DOUBLE SCOTCH"

Prestwick's Pioneers

"Pioneer" - the name was aptly chosen, not only because the manufacture of an entire aircraft to its own design broke new ground for Scottish Aviation, but also as a description of the role which the aircraft was intended to perform.
The Prestwick Pioneer was designed in 1946 to meet an unusual specification based on a combined Army, Navy and Air Force requirement for a front-line communication aircraft which would be able to carry four passengers and pilot and to operate from rough surfaces and small landing grounds not exceeding one hundred yards in length. The specification therefore called for steep angles of approach and climb in order that the aircraft could be operated in very confined spaces. It had to be an all-metal construction capable of withstanding long periods in the open and of being maintained by semi-skilled personnel isolated from technical facilities and in severe climatic circumstances. In short, the aircraft had to be able to operate in pioneering conditions.

The Prestwick Pioneer could not, however, claim to be the first aircraft with S.T.O.L. (short take-off and landing) capability. Its precursors stretched back through the Fieseler Storch, built in Germany during World War II, to the 1920's when experimental work on various high-lift devices culminated in the Guggenheim Safe Aircraft Competition. Staged in America in 1929 this contest required entrants to be able to achieve a top speed of at least 110 mph, a minimum speed of 35 mph, the ability to land in one hundred feet and the ability to take off in three hundred.

At least two of the twenty-seven entries, the Handley Page Gugnunc and the eventual winner, the Curtiss Tanager, achieved the performance specification. But both these aircraft were slotted bi-planes and by the time that Scottish Aviation was designing the Pioneer it seemed as if the obvious answer to the requirement to operate in and out of confined spaces had been provided by the helicopter. The Company's own largely unsuccessful efforts to sell Bell helicopters under licence had, however, coloured its views on the subject of rotary wing machines and David McIntyre liked to describe the twin engined development of the Prestwick Pioneer which the Company undertook in 1953 as the logical answer to the public's "fanciful expectations from helicopters".

The object of scaling up the dramatic take-off and landing characteristics of the Prestwick Pioneer to produce the Twin Pioneer was to enable a commercial load of up to sixteen passengers to be carried in areas where either geographical conditions made it impractical or where the traffic density or economic situation did not justify the capital expenditure necessary for airfields to be constructed. It was in fact the terrain and conditions of Scotland's own Highlands and Islands which served as the initial model for the aircraft's design.

Twin Pioneer operating from the frozen lake at St. Moritz, Switzerland.

Sunnier climes were also used to illustrate the advantages which STOL aircraft had to offer in terms of comparative runway costs. A comparison made on two islands in the West Indies demonstrated that, whereas the costs of establishing runways for a Viscount aircraft were £650,000 and £345,000 respectively, it would only require £45,000 to build a strip for the Twin Pioneer.

But it was not simply in the frontier territories, geographic or economic, that the Company expected the Twin Pioneer to prove to be a winner. It was also envisaged that the aircraft could serve very successfully as a short range inter-city vehicle because of its ability to operate from much closer to city centres than conventional commercial aircraft.

In October 1955 Scottish Aviation was contemplating the introduction of a London-Paris service using the aircraft, having established that the overall journey between the city centres was four hours by Viscount, while the Twin Pioneer, despite its slower cruising speed, could cut this time by an hour and a half and the fare by a third. On the other hand the proposed service perhaps lost some of its passenger appeal as a result of the fact that while the French terminus was at the Champs Elysees, the embarkation point in London was to be Wormwood Scrubs.

Other applications for which both the Prestwick Pioneer and the Twin Pioneer were expected to be particularly suitable included aerial photography, remote air mail deliveries, crop dusting, casualty evacuation and airborne broadcasting. In the air ambulance role the Pioneer was configured to carry two stretcher cases and medical attendant, while the Twin could accommodate nine stretcher cases, two "walking wounded" and two attendants.

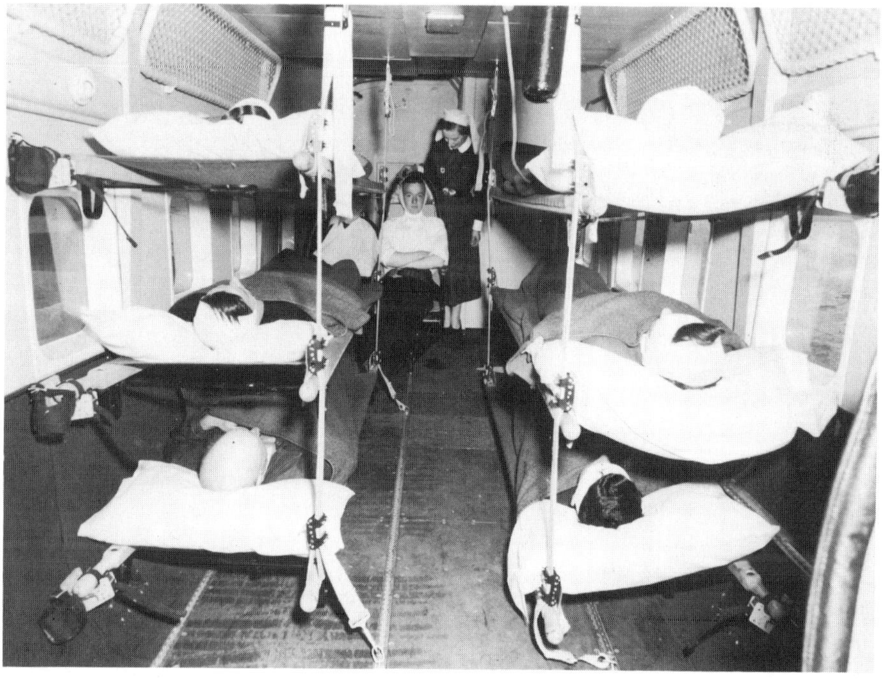

Casualty evacuation.

In Philippines Air Lines configuration

The principles governing the design of the Pioneers were, as the Company's sales brochures explained, "high lift techniques which have been in use throughout the whole life of aviation and which have been proved by nature over countless ages in the evolution of large cliff-dwelling birds". The original brochures proceeded to become somewhat carried away by this rhetoric, adopting the slogan
"Fast through the air - slow and precise near the ground"

With an all-out speed of only 124 mph from its original De Havilland Gipsy Queen 34 engine the only charitable conclusion would be to assume that the brochures meant that the aircraft was fast through the air by comparison with the cliff-dwelling birds from which its design had been derived.
It was the second claim which was the heart of the design intent and with a stalling speed of only 33 mph and the photographed ability to land in a mere 26 yards the Prestwick Pioneer certainly achieved this intention.
The Twin Pioneer in its turn staggered spectators at Farnborough when it demonstrated that it was possible for a twin-engined 16-seater to land in only 60 yards.

The Prestwick Pioneer was an all metal structure although the primary control surfaces were fabric covered. Appropriately for the Company run by David Fowler McIntyre, the aircraft's slow flying qualities were achieved through the use of large area Fowler flaps and

full spar leading-edge slats which were 25% chord surfaces.
Relative to the size of the aircraft these were the largest ever known to have been built when Flight magazine appraised the Pioneer in October 1947. The same review concluded that the Prestwick Pioneer was, for its weight and size, perhaps the slowest flying aircraft in existence.

Prestwick Pioneer prototype with original Gypsy Queen 34 engine.

The other principal significant feature of the design also sprang from this slow-flying capability which required the aircraft to take a severe nose-up attitude to achieve the correct angle of attack for landing. In order to avoid a tailplane stall in these circumstances the design provided for the tailplane to be pivoted on an axis close to its leading edge, so that its incidence could be negatived by means of an electric screw jack.

Prestwick Pioneer production line with Scottish Airlines' Walrus Flying Boat.

The adjustable angle of tailplane movement was eventually established within the range +4° to -9°. The high nose-up attitude was also responsible for the Pioneer's tall undercarriage, which in turn made it necessary to hang a stirrup from the fuselage and called for a certain personal agility to enter and leave the aircraft.

Whatever the eccentricities of its profile, 'Flight' pointed out that on the score of workmanship the prototype Prestwick Pioneer was the most beautifully built aircraft which it had examined for some considerable time, a feature which was all the more notable in view of the particularly light, 26 gauge, material used for the skin panelling.

The Twin Pioneer for its part was aimed to provide inexpensive air travel on journey sectors of up to 500 miles. The wide, almost square section fuselage was designed to offer scope for either a high degree of passenger comfort or alternatively maximum space for freight.

The cabin was nineteen feet long in the standard aircraft with 5' 8" headroom throughout its length. Large, rectangular windows and the high wing arrangement gave excellent fields of vision for the passengers.

The wings of the Twin Pioneer were a simple development of its single-engined predecessor again utilising the large Fowler flaps. In order to reduce tooling costs the wing panels of the Prestwick Pioneer were incorporated directly into the design of the Twin.

There were those who criticised the appearance of the Twin. During a sales tour in the U.S.A. one Air Traffic Controller demanded to know if the pilot had made the aircraft himself before the control tower would allow it to land. Others suggested that its high strutted wings and box-like shape were old-fashioned, "like a 1930's aircraft". The "Groupie" responded by saying that was exactly what he wanted it to resemble.

Perhaps the two most memorable characteristics of the Twin Pioneer's appearance were its tall, non-retractable undercarriage and its triple tail fins and rudders. The undercarriage arrangement with its tail wheel derived from the intention that the aircraft should operate from remote, roughly prepared airstrips on which there would be a high risk of a tricycle undercarriage aircraft "nosing over" when braking. Designed as a slow-flying aircraft in the first place the extra drag resulting from the fixed undercarriage was considered to be comparatively insignificant.

The original design envisaged a single vertical stabiliser. The Twin Pioneer's triple tail sprang from the necessity for a twin-engined aircraft approaching with an engine failure to be able to overshoot in the event of an aborted landing. Using full power on one engine for this purpose obviously created an asymmetric thrust which needed to be counteracted by additional control. The Twin's large tail was designed specifically to give this extra control.

The 'Flight' reviewer who flew the Twin Pioneer in 1955 found that the aircraft had the most remarkable stalling characteristics in which it seemed to give a "slight shrug" before simply continuing in its glide with its wings a few degrees lower but still level. In this respect too the aircraft had the same good nature as the Prestwick Pioneer, even though the Twin had a significantly higher stalling speed than its predecessor, nearly 48 mph.

'Flight' also reviewed the Twin Pioneer Series 3 five years later, following the installation of the Alvis Leonides 531/8 long stroke, direct injection engines and was favourably impressed, not only with the improved single engine performance but with the steady, solid feel of the aircraft which, like the Dakota, inspired tremendous confidence in its passengers. On one occasion when a Twin Pioneer crashed in Borneo, not only were none of the passengers hurt but a baby sleeping in a box in the rear of the aircraft was not even woken up in the process.

..."the most beautifully built aircraft".....

The Chief Designer responsible for both the Prestwick Pioneer and the Twin Pioneer was Robert McIntyre, who had worked with Shorts of Rochester and then with Hawker Aircraft at Kingston before coming to Scottish Aviation in 1942 with twenty years of experience in the industry.

Handover of first Prestwick Pioneer to the R.A.F.

Official Roll-Out of G-ANTP

Prestwick was not Kingston, as he was to discover when both Pioneers came to be developed with no wind tunnel tests whatsoever, despite the unprecedented size of the flaps. Nevertheless, both aircraft unquestionably achieved the short take-off and landing abilities for which they were designed, much to the credit of T. B. Lyon, who was then Chief Aerodynamicist. On one occasion Robert McIntyre received a triumphant telephone call from the Royal Aircraft Establishment at Farnborough announcing that the Pioneer which it was testing had surpassed itself by managing a forced-landing on a cricket pitch! Unfortunately even the Pioneer was incapable of taking off again from this strip and the aircraft had to be dismantled to get it back to Farnborough.

The Pioneer's slow-flying speed quickly became a local phenomenon around Prestwick. Motorists driving towards the airport on the Kilmarnock road were frequently startled to find themselves overtaking the aircraft! On one occasion Robert McIntyre and Chief Test Pilot Noel Capper were making their approach to the airport in the Pioneer only to be asked by Air Traffic Control to "stay put" while a C54 took off. The nose wind was such that they obliged by following this request to the letter and were later asked by one onlooker to confirm that he had actually seen the Pioneer "pinned to the sky".

On other occasions observers commented that the aircraft could be seen literally to stop when it was throttled back and in strong winds even to be flying backwards!

Part of the Company's boast for the Twin Pioneer was that it was able to take off at the speed of a pedal bicycle. One disbeliever saw fit to challenge this assertion and so a demonstration was duly arranged with aircraft and bicycle setting off simultaneously from a standing start. The cyclist was in fact in the lead when the aircraft left the ground and, even though the organisers had taken the precaution of employing a sprint cycling specialist, both the bicyclist and the aircraft emerged from the demonstration with distinction.

Landing in 26 yards.

Among the many interested visitors who came to see these remarkable aircraft which were being built at Prestwick, was the Duke of Edinburgh who visited the factory on 13th October 1954 and who, besides flying in the Pioneer, made his own unique contribution to the Company's history by suggesting that perhaps the Twin Pioneer should have been christened the "Double Scotch".

But the Pioneers were more than merely curiosities and the success of the Scottish design in practice was most forcefully illustrated by the aircraft's achievement in service with the Royal Air Force in Malaya. Over Christmas 1954 and New Year 1955 David McIntyre made a three week tour to gain a first-hand impression of the aircraft's performance in this theatre and to assess the market for the Pioneer and the Twin in the Far East.

It is evident from his journal of this tour that the RAF, which had received its first Prestwick Pioneers in Malaya in 1953, was an enthusiastic user.

> "It became clear" he wrote, "that they reckoned that they had struck a winner with the Pioneer, that their new methods of jungle warfare depended on it, and that they had grave fears of our ability to produce Pioneers in the quantities required. It was also clear that the Pioneer has had a remarkable record of serviceability and a negligible spares requirement and that it was standing up nobly to intensive use and very rough treatment".

The Pioneer's role in Malaya was to provide support to the garrisons which were being established throughout the jungle to secure the surrounding areas from the threat of communist infiltration. In addition to their own communications and supplies role the Pioneers had the job of retrieving the parachutes by which RAF Valettas had dropped bulk supplies.

The landing strips used by the Prestwick Pioneers were typically rough clearings in mountainous terrain covered in dense jungle with trees up to 300 feet high. Occasionally they were created by levelling the tops of small hills, as at Fort Dixon where the strip was 180 yards long, with precipices on all sides, "rather like a tiny aircraft carrier", as McIntyre observed.

At Fort Shean, on the other hand, landings took the aircraft to within ten yards of the river bank. Landing here was seldom into wind and was usually complicated by cross-winds which were constantly changing as a result of the heat turbulence of the valleys.

A typical airstrip in the Malayan jungle, Fort Shean

Among the other hazards were the torrential rain showers which could turn bare ground into a muddy skating rink and which therefore made it necessary for there to be a suitable grass surface. However, this in turn created its own problems and two Malays with long-handled sickles had the task of keeping the grass to exactly the right length, since if it grew too high it made the tyres skid on the flattened grass blades and caused an extra drag during take off. The effect was marginal, but so were the limits to which the aircraft was operating on these fields. Another difficulty to be faced was the bamboo which grew on the approaches to the airstrip and which had to be kept carefully trimmed since its vigorous rate of growth in the Malayan climate was such that the margins available for a safe approach could be the equivalent of no more than one day's growth.

Landing at Fort Shean took the aircraft to within 10 yards of the river bank

Yet the most dramatic of the various landing strips which David McIntyre visited during his trip was that at Fort Legap. His own account conveys just what it was like to fly into these primitive airfields.

> "Legap was spotted at the bottom of a very narrow valley, which was more like a gorge; then we flew about nine miles east to enter the planned approach, which is a tortuous route through descending valleys between the mountains, at times with only yards to spare from the wing tips on either side. The strip is out of sight throughout the approach, until the very end when a sharp left-hand turn is made round a large tree, which is rather a darker shade of green than the rest, and immediately in front is a river, a 200 yard strip and at its end a sheer cliff face and a 1,500 foot mountain. A further five yards has been laboriously dug out of the face of the cliff to provide a little additional run which might be required if the grass surface has not been cut to the correct length".

If landings in these circumstances could more appropriately be described as "controlled drops", then take-offs were scarcely less exciting and were only made possible by the aircraft's very steep angle of ascent and its negligible forward movement during climb. It was not long before the RAF Pioneer pilots had christened this manoeuvre "the clamber out".

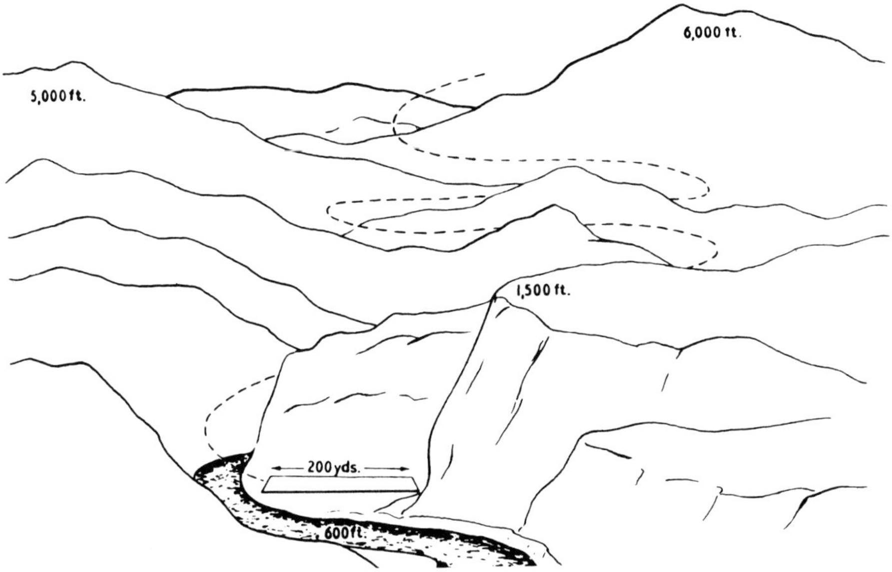

The approach to Fort Legap

A measure of the flying skill needed in these circumstances is provided by the fact that the RAF stipulated a minimum requirement of six months training for selected pilots before they would be considered capable of undertaking Pioneer operations in Malaya.

The forts which the Prestwick Pioneer were supporting typically consisted of a garrison of between 40 and 75 Malayan police with two European police officers, ordinarily of the rank of Lieutenant. Accommodation took the form of huts which the local Malays put up in a few days surrounded by barbed wire and often with a flock of geese to serve as watch dogs. While one half of the garrison stood by the fort the other had the job of policing the surrounding jungle, which involved not simply tracking down terrorists, but making and maintaining contacts with the local population, many of whom were brought on to the ration strength of the fort and paid to take part in its work.

In addition to providing medical centres the forts became nuclei for small shops selling items such as shirts and shorts, skirts, lipstick, hairpins and torches and buying what was at the time of David McIntyre's visit the natives' only exportable product - blow pipes!

The extent to which this can be described as a process of

"civilisation" obviously involves value judgements and must be a matter of opinion but to David McIntyre the Malayan experience seemed to be tangible evidence of his deeply-held belief in the civilising effect of the aeroplane and the unique capability of the Prestwick Pioneer in this respect.

The Pioneer Flight in Malaya was based at Kuala Lumpur. "I have never met", his journal records, "a more enthusiastic RAF unit".

> "They were obviously enjoying themselves up to the limit in the knowledge that they had a big and important job to do, and I think, like me, they have a feeling that they are making aviation history".

Certainly the pilot of one Prestwick Pioneer operating in Borneo must have a fair claim to having made history in a flight which carried to safety a mother, father and fourteen children trapped by rising floodwaters.

Subsequent Pioneer and Twin Pioneer sales brochures pictured the aircraft operating not only in the jungles of Malaya but in conditions as different as the Himalayan foothills of India, the frozen lakes of Switzerland and Canada, the beaches of the Bahamas and even a polo field in Teheran. Early admirers of the aircraft ranged from the Shah of Iran to the headhunters of New Guinea.

Operating from the beach at Harbour Island in the Bahamas

At Santa Rita in the Andean mountains of Colombia, a community one and a half day's pony ride from the nearest road, the villagers had carved out a landing strip for single-engined aircraft. When a Twin Pioneer demonstrator landed there the local Priest rushed down with a bottle of champagne to celebrate an innovation which could mean so much for the development of the area.

In the backwoods of Borneo most of the landing strips were short, narrow affairs which had been cut out for the single-engined aircraft used by missionaries. Typically these strips were no more than fifteen yards wide, considerably narrower than the Twin Pioneer's 76 foot 6 inch span, which left the wing tips barely above the adjacent undergrowth. What the Twin Pioneer had to offer inaccessible territories like Borneo was the opportunity to market crops, such as tobacco, coffee and pepper which grew readily in the jungle climate, swiftly and in bulk.

At the other extreme the Twin Pioneer demonstrated its peculiar capability for short range inter-city travel by operating during the Paris Air Show in 1957, between Le Bourget and the Paris Heliport at Issy-les-Moulineaux near the Eiffel Tower.

The message carried by Scottish Aviation's sales brochures was simple:

> "There is no territory which cannot have Pioneer air strips and the countless benefits of air transport simply and cheaply. There is no city which cannot have a Pioneer air strip and air transport on its doorstep instead of at some distant airport".

Perhaps the most interesting single revelation of David McIntyre's journal of his Far East tour is that before visiting Malaya he had lost his confidence that there was a market for the Prestwick Pioneer. By the time he returned these doubts had gone and he was confident not only that there was an increasing and lasting market for the Pioneer in the development of colonial and other territories but that there was a need for Twin Pioneer output to increase

> "well beyond the capacity of our own organisation".

It certainly seemed as if the aircraft had the potential to provide further evidence of the aptness of Scottish Aviation's motto, "The World O'er". Flight's review in 1955 had concluded that the versatility and economy of the Twin Pioneer had scarcely started to be appreciated and that the aircraft was in a class of its own, with world-wide appeal.

As early as September 1954 the Duke of Hamilton was able to report to his fellow Directors that during a recent visit to the United States several leading American aircraft manufacturers were already expressing interest in the Twin Pioneer and in the possibility of manufacturing the aircraft in America under licence, even though the aircraft was not to make its first flight until 24th June 1955.

In 67½°of frost at Seven Islands, Canada

By May 1955 the Malayan Auxiliary Air Force was suggesting that it might wish to place an order for 40 Prestwick Pioneers and by July there were indications that the Royal Air Force was considering a requirement for at least 50 and possibly 100 Twins. The Auxiliary Air Forces of Borneo and Hong Kong looked like other likely buyers for the Pioneer with a view to employing it in anti-piracy roles. The European Sales tour had resulted in the first commercial Pioneer sale to the Austrian Air Rescue Service and had revealed a possible military order for between 40 and 50 aircraft for Germany. India and France both offered the prospects of substantial orders and a sales tour had been despatched to Africa.

David McIntyre grew increasingly anxious to purchase three Twin Pioneers for Scottish Airlines, not only to take advantage of the opportunity to demonstrate the aircraft in operation but also because he was concerned that, if he did not get them quickly the demand from external customers would be so great that Scottish Airlines would have to wait a very long time to get any.

The Austrian Air Rescue Service with their Prestwick pioneer.

Against this background the plans which were made seemed at the time to be relatively cautious. The aim was to manufacture 100 Pioneers and an initial batch of 200 Twin Pioneers.

As the Company reached its twenty-first birthday in 1956 it seemed literally to have come of age. There was evidence of strong market interest in the aircraft which it had designed and employment was building up rapidly. The Twin Pioneer programme was reckoned to require an increase in production operators at Prestwick from an existing level of 670 to a new total of 1500 and by the middle of the year the Company's various activities employed nearly 3,000.

The difficulty of recruiting sufficient skilled labour at this time was in fact a major constraint and as a result the centre section of the Twin Pioneer was sub-contracted to Armstrong Whitworth Aircraft.

Building aircraft to its own design the Company's morale was high. The General Manager was even able to report to the Board "that the most cordial relationship existed between the Factory and the Design Department in connection with the Twin Pioneer development".

Above all this morale was sustained by David McIntyre, "The Groupie". He was reputed not simply to know everyone in the factory by name but even to know the names of their children. He used to invite the staff to sports days and bonfire nights on the grand scale in the grounds of his home, first at the Cushats near Coylton and then at Lochgreen in Troon. On one occasion a lighted rocket failed to take off as it should and landed in an open box containing over £30 worth of fireworks, a veritable arsenal in those days. The results were spectacular and absolutely characteristic of a man whose favourite colour was flame-orange and with whom spontaneous acts of personal generosity came naturally.

He was the man who called in with hot thermos flasks and sandwiches when his men were working late in the Flight Shed or who arranged for employees and their families to go on holiday to Belgium if they had worked, to respond to some Company requirement, through the factory's normal summer holiday shutdown.

On one occasion McIntyre suddenly announced to the Superintendent of the Experimental Shop, who had been responsible for working all hours on the loft layouts for the Twin Pioneer, that he would give him a holiday anywhere in the world and asked him where he wanted to go.

The Superintendent, thinking that this was probably a joke, replied in kind by saying that having dealt with the loft-lines and jigs for the Twin Pioneer, he would be curious to see the Great Pyramid of Cheops in Egypt, because he had heard that its base was nine minutes out of alignment on one side and he wanted to check it. Shortly afterwards he found himself flying to Cairo on a Scottish Airlines York together with his theodolite! He returned with the position of the pyramid clarified - it was indeed a few millimetres out of line - but was never able to solve the altogether different mystery of why, when walking to the pyramid, one of a group of crouching Arabs should suddenly call out to him "Hey, are you fae 'Mulguy', Jock?", although he might have been forgiven for supposing that it was David McIntyre who had somehow conjured this unlikely genie out of the sands.

The Pioneer period was one when, neither for the first time nor the last, Scottish Aviation financially had need of a conjuror. Although the factory had been brought back into profit and the Company had achieved a substantial award in settlement of its claim for compensation following the compulsory acquisition of Prestwick Airport, the scale of the investment required to launch the Twin Pioneer project in 1954 was far beyond Scottish Aviation's own means.

Consequently when interest was first shown in the idea of a twin-engined Pioneer in 1952, the Board decided that if the aircraft were to go into production it would have to be under a Government Development contract.

The Ministry of Supply for its part liked the

specification of the aircraft but was not prepared to commit itself until the end of 1953 and even then was reluctant to provide more than one-third of the costs of the Twin Pioneer prototype.

Twin Pioneer prototype emerging from T.I. Hangar

An approach was therefore made in the middle of 1954 to the Finance Corporation for Industry. The FCI was a Government-established body whose role was the provision of finance for industrial projects in which the risk was reckoned to be too high for loans to be obtained, at viable rates of interest, from banks or other normal commercial sources.

On 26 October 1954 the FCI undertook to grant the Company an unsecured loan of £1,000,000 for the development and production of the Twin Pioneer. The decision was subject not only to the consent of the Treasury and other relevant Government Departments but also to the Company securing a 99-year lease of its factory area at Prestwick Airport and obtaining the £50,000 Development Contract from the Ministry of Supply.

Having satisfied these requirements there were a number of other conditions on the loan, notably that the Company's authorised capital was not to exceed £600,000 without FCI's consent and that FCI should have the right to acquire up to £200,000 ordinary shares in the Company at par at any time before 31 December 1961. These additional terms also precluded Scottish Aviation either investing further money in any of its subsidiaries or granting credit or making loans to them, particularly the airline operations, without the consent of the FCI.

In addition, the Finance Corporation for Industry acquired the right to appoint two Directors and it was in this way that Major Jackson Millar and H W Fulton, who were respectively Chairman and Managing Director of Albion Motors, joined the Scottish Aviation Board. Their presence was ironic in that SAL had only recently cancelled its Albion Bus programme, but their broad industrial experience was welcomed, as the magnitude of the task in front of the Company quickly became apparent.

A novel incentive scheme was introduced at this time under which members of the Management Committee took it in turns each week to broadcast details of the project's progress across the tannoy. Employees were encouraged to listen carefully by the prospect of winning a bottle of whisky for correctly identifying the anonymous voice.

In the summer of 1954 it was clear that the Twin Pioneer would not be ready for that year's Farnborough, although it was hoped that the prototype would be complete within a couple of months after the Exhibition. In the event G-ANTP did not fly until June the following year, twenty-one months and 93,000 design manhours after the commencement of the design.

Officially the Twin Pioneer first flew on 25 June 1955, but in fact the aircraft had its wheels off the ground, unofficially, the previous day with a squad of enthusiastic fitters in the cabin for ballast. After the long months of preparation this anxiety to see the aircraft airborne was understandable and the aircraft itself seemed keen to get on with the job since it was only by shouting instructions to his human ballast that the pilot was able to coax it back on to the ground.

But first flight, as ever, was the beginning rather than the end of the development phase. Hopes that the final Certificate of Airworthiness would be obtained by the Spring of 1956 proved to be much too optimistic. Type certification was eventually achieved in November 1956 and it was just over two years to the day after first flight before the first customer deliveries of production aircraft were made. These were three aircraft for the KLM subsidiary De Kroonduif NV, aircraft numbers 509, 510 and 511 which left Prestwick together on 26 June 1957 for delivery to Dutch New Guinea.

The first four aircraft, 501 - 504, had all been prototypes and demonstrators. 504 (G-AOEP) achieved the distinction of being the first Twin Pioneer to be sold when it was purchased by the Zinc Corporation Limited of

Australia on 1 May 1957 after the completion of its Far East and Australian tour. 505 was the first production aircraft and made its maiden flight on 13 February 1957, but was not sold until 15 September 1959, when it went to the Rio Tinto Finance and Exploration Company equipped as a geophysical survey aircraft. 506 became the static test airframe while 507 and 508 were ordered by BOAC Associated Companies. They were originally intended for Borneo Airways, but went instead on charter to the Kuwait Oil Company. When these aircraft left Prestwick at the end of October 1957 a total of twelve Twin Pioneers had flown but only half of these had been sold.

Although interest in the aircraft was widespread from the beginning, firm orders were slower to materialise than had been anticipated and this, coupled with the project's extended design and development phase, meant that costs and work-in-progress soon escalated beyond the original expectations. The Company's requirements for additional finance consequently mushroomed.

Fortunately the Finance Corporation for Industry was sufficiently confident in the aircraft's potential to be prepared to inject further funds. An additional £250,000 was made available in July 1956 but it quickly became apparent that this was not adequate. By the end of March 1957 the FCI had advanced another £500,000 in return for a 45% rather than a 33% holding of Scottish Aviation's share capital. The total loan was approaching £2½ million by the time that the first aircraft deliveries were being made in the summer of 1957 and in an attempt to control the escalation of the project's cash demands the decision was made in July that the mature production rate should be cut from its originally planned level of eight per month to the still ambitious target of six.

First customer aircraft

Against this background the Twin Pioneer sales tours acquired a new urgency. As usual David McIntyre threw himself into the task but on this occasion, as fate would have it, "the Groupie" stepped over the line which divided the energetic from the impulsive. On 7 December 1957, G-AOEO, which had embarked on a demonstration tour of North Africa, crashed in the Libyan desert near Tripoli. Captain R Smith, Flight Engineer R C Clapham and David McIntyre were all killed.

The cause of the accident was quickly traced to a fatigue failure in the main strut's fitting to the port wing and the collapse of the front Vee brace tube. The tragedy was that the reliability of the T2 tube material of the fitting had already been identified as a potential problem, T50 had been specified as the preferred alternative and action had been taken to replace the T2 material with T50. G-AOEO was in fact the only aircraft which had not had this modification embodied by the time that the North African demonstration tour was due to start. A number of David McIntyre's colleagues urged him to postpone the tour until the modification had been carried out, but he was not prepared to accept the delay. The job would have taken a mere two days to complete.

With the benefit of hindsight, the decision to proceed on the tour with the unmodified aircraft seems to have been an unnecessary gamble. At the time the risk must have appeared slight and it should be remembered that this accident was one of the earliest recognised examples of metal fatigue, a problem which was only thrown into prominence by the Comet disaster. Scottish Aviation for its part had dedicated a Twin Pioneer to a two year programme of airframe fatigue test. When G-AOEO crashed it had completed only 564 flying hours, whereas tests had given the T2 material a safe life of over 2000 hours.

What had not been taken into account was the fact that the original calculations to establish the safe life were based on a pattern of utilisation in which the aircraft would take off, climb to 3,000 feet, cruise at that altitude and land. As a demonstrator EO was operating to a very different pattern in which it was taking off and landing much more frequently and was seldom cruising any higher than 1000 feet where it was liable to the additional stresses imposed by the greater turbulence experienced in low altitude flying. Coupled with this, the aircraft was operating in the climatic extremes of the Libyan desert.

When the fatigue calculations were re-run to simulate the actual conditions in which G-AOEO had been operating when it crashed the results showed the metal fracturing after as little as two hours flying.

David McIntyre's death was a shattering blow to the Company's morale. Appropriately it was his fellow conqueror of Everest and co-founder of Scottish Aviation, Douglas, Duke of Hamilton, who recorded the Company's sense of loss.

> "Group Captain McIntyre always played the prominent part in the inception and meteoric development of the Company and the Airport at Prestwick before and during the war years. On him fell the organisation and management of the first Atlantic air service from this end and, during his nine months in control, not a single serious accident occurred. Since the war he struggled, sometimes against great opposition, for the establishment of an aircraft industry in Scotland, which is now an accomplished fact with an approved design organisation. He was a man of vision and dynamic personality and his death in the prime of his life is a sad loss not only to the Company but to Scotland and Aviation".

Unquestionably David McIntyre also had his critics and his opponents. No one who pursued any cause with the relentless determination and single-minded energy which he consistently displayed could avoid doing so. The fact that he was seeking to establish and develop in Scotland an industry which has always been particularly subject to political pressures made this doubly certain.

On the other hand, it was largely those same personal qualities which explain the commitment that he secured from those who worked within the Company. He had immense qualities of leadership and when he died the Company lost the man who had always been its personification.

The fact that he died in the aircraft which was carrying Scottish Aviation's future hopes and aspirations made the blow all the more severe. Quite apart from the impact on the Company's morale the immediate effect of the Libyan crash was of course to damage market confidence in the Twin Pioneer badly. Prospective orders, including prestigious customers such as British European Airways, evaporated while existing buyers asked for delivery of their aircraft to be deferred and deposits to be returned. This process was only partially offset by the announcement of an additional order for 12 aircraft for the RAF.

The Company responded by a further intensification of its sales efforts during 1958 including demonstration tours to the Middle East and South America. It was an uphill struggle. By this stage the FCI's loan had grown to £3

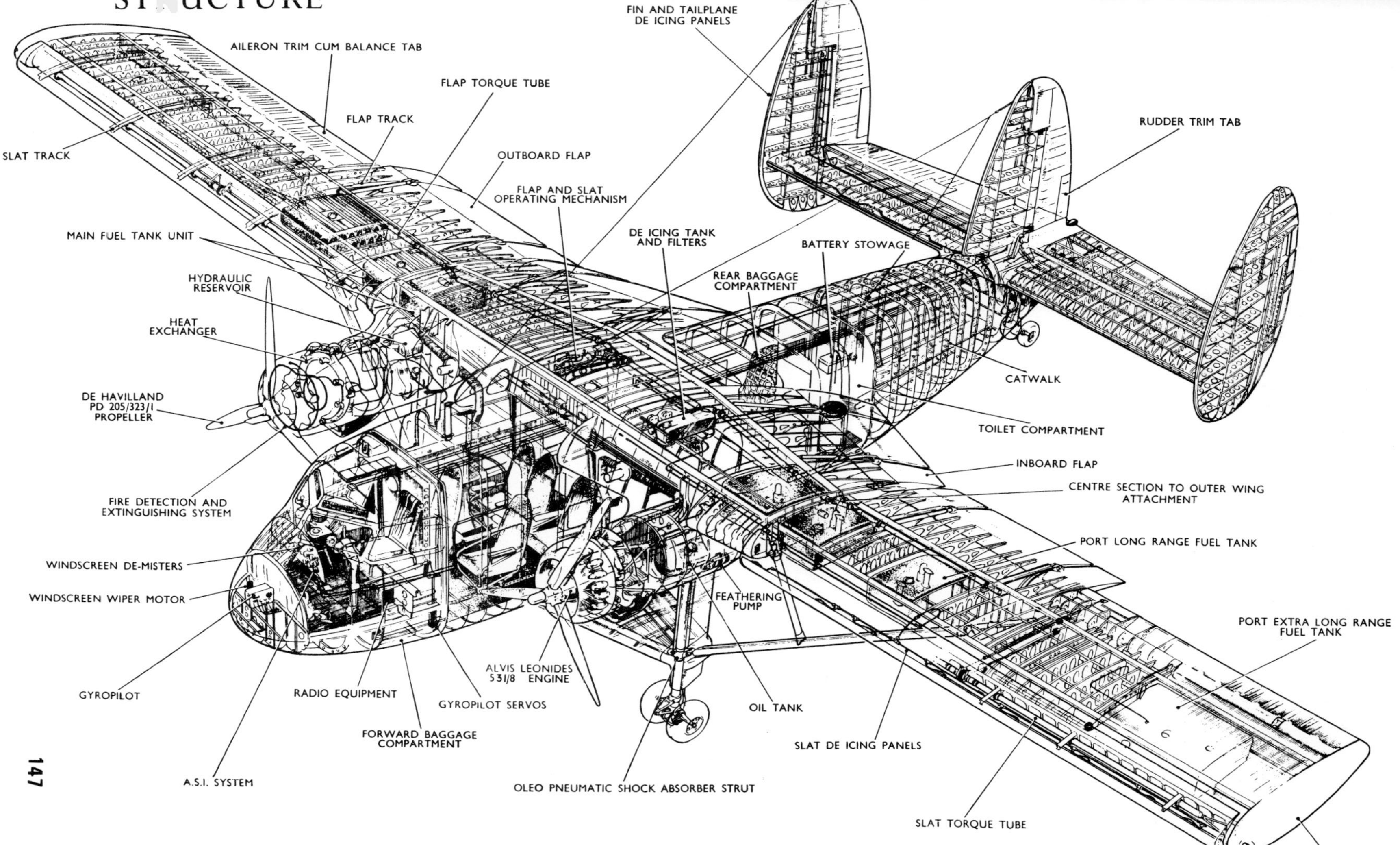

Twin Pioneer structural diagram

million and the Company's financial position had deteriorated to such an extent that it was decided even before the start of the South American tour that it would be too expensive to bring the demonstrator back to the United Kingdom and that it would have to be sold in Mexico.

1958 was the peak year of Twin Pioneer production with a total of 37 aircraft coming off the production line. Reflecting the impact of the Libyan disaster, however, the rate was progressively cut back during the year, initially to four aircraft per month in March and then to two per month in July.

Ironically the aircraft was one which pilots found easy to handle and in which passengers felt particularly safe. Unfortunately the G-AOEO crash in Libya was not the only accident to mar the aircraft's reputation.

The incident which had first drawn attention to the possible weakness of the wing strut had occurred on 30 August 1957 when one of the de Kroonduif Airlines Twin Pioneers, JZ-PPX, had crashed off the Biak Coast in New Guinea. At the time it had only completed 250 flying hours, the crew were all killed and the aircraft was not recovered. It was therefore assumed at first that the accident had been the result of pilot error. The only witness to the incident was a native fisherman who reported that a wing had fallen off, an observation which led directly to the investigation of the T2 strut fitting material and its replacement.

The Libyan crash came little more than three months after the loss of JZ-PPX and the proximity of the two accidents in the early stages of aircraft deliveries was particularly damaging to the Twin Pioneer's prospects.

Tragedy struck again when G-ANPP, the first prototype, which had by now been converted into a Series 3 aircraft, crashed at Jorhat in Assam on 10 August 1960, while being demonstrated to Indian Airlines and with an Indian Air Force pilot at the controls. The cause in this case was reckoned to be pilot error, but this did not mitigate the blow.

Only a few months earlier, on 12 December 1959, G-AOEN, the second prototype, also converted to Series 3 standard, had made a forced landing on an island in the middle of the Zambesi River in Mozambique. Although none of the passengers were seriously hurt the aircraft was a total loss. Yet even in these bleak circumstances there were those who went on looking on the bright side. Perhaps it reflected the resilience of Scottish Aviation personnel or perhaps it was simply the eternal optimism of the aircraft salesman, but the suggestion was made that the Zambesi accident might actually help further sales in Mozambique because the crashed aircraft would be a ready source of cheap spares!

There were a number of reasons why neither the Prestwick Pioneer nor, more especially, the Twin Pioneer fully realised the original expectations which had been held of them. The early crashes, particularly the two resulting from the failure of the wing strut fitting, have made it easy to conclude that the principal problem was a technical one, but this affords only a partial explanation.

The Twin Pioneer fatigue test rig

"A ready source of cheap spares", Mozambique

Certainly the original Prestwick Pioneer was under-powered. This problem was addressed when the Gipsy Queen 240 hp power-plant, called for by the original Ministry Specification, was replaced with the Alvis Leonides 501/3 rated at 520 hp and then with the 502/4 560 hp engine. This increased the maximum speed from 114 mph to 162 mph and the payload from 680 lbs to 1,200 lbs over a range of approximately 350 miles.

Performance Summary of Prestwick Pioneer with Alvis Leonides 502/4 engine.

GENERAL DIMENSIONS

Length	45′ 3″	(13 m, 8)
Span	76′ 6″	(23 m, 32)
Height (Tail Down)	12′ 3′	(3 m, 7)
Wing Area (including ailerons)	670 sq. ft.	(61 m² 3)
Wing T/C at Root	15%	
Wing Mean T/C	15%	
Wing T/C at Tip	15%	
Wing Section—Root	N.A.C.A. 4415	
Wing Section—Tip	N.A.C.A. 4415	
Incidence of Wing Chord to Fuselage Datum	4°	
Tail Plane and Elevator Area	167.2 sq. ft.	(15 m² 53)
Tail Plane and Elevator T/C	12%	
Fin and Rudder Area	167.3 sq. ft.	(15 m² 54)
Fin and Rudder—Mean Thickness	9%	
Flap % of Wing Chord	30%	
Slat Outer Wing % of Wing Chord	25.1%	
Fixed Slat, Centre Section % of Wing Chord	14.75%	
Flap % of Wing Span	50.7%	
Flap, Angular Range—Take-off	14½°	
Flap, Angular Range—Landing	32½°	
Type of Flap	Fowler	
Fuselage Datum Angle to Ground at Rest	11½°	

Performance Summary of Twin Pioneer with Alvis Leonides 503/8 engines.

PERFORMANCE SUMMARY AT 13,500 lb. (6123 kg, 4)
in
INTERNATIONAL STANDARD ATMOSPHERE

Maximum speed	180 m.p.h. (289,72 km h) @ 1,750 ft. (533 m, 4)
Maximum Continuous Cruising Speed	174 m.p.h. (280,07 km h) @ 7,250 ft. (2209 m, 8)
Maximum Weak Mixture Cruising Speed	159 m.p.h. (255,83 km h) @ 11,000 ft. (3352 m, 8)
Cruising Speed at 5,000 ft. (1524 m) on 40% Take-off Power	124 m.p.h. (199,52 km h)
Specific Range at 5,000 ft. (1524 m) on 40% Take-off Power	4.13 m.p.g. (1,462 km/L)
Cruising Speed at 5,000 ft. (1524 m) on 50% Take-off Power	139 m.p.h. (223,65 km h)
Specific Range at 5,000 ft. (1524 m) on 50% Take-off Power	3.76 m.p.g. (1,33 km L)
Stalling Speeds—Flaps up Engine off	67.5 m.p.h. (107,80 km h)
Flap 14½° Engine off	51.5 m.p.h. (82,86 km h)
Flap 32½° Engine off	47.5 m.p.h. (76,43 km h)
Maximum still air range at 124 m.p.h. (199,52 km h) at 5,000 ft. (1524 m) with full tanks ...	620 miles (997 km, 6)
Maximum still air range at 139 m.p.h. (223,65 km h) at 5,000 ft. (1524 m) with full tanks ...	565 miles (909 km)
Payload under the above conditions	3,008 lb. (1364 kg, 4)
Rate of climb at sea level	880 ft. per min. (4,4 m s)
Take-off distance to 50 ft. (15 m, 24)—concrete runway ...	342 yds. (312 m, 72)
,, ,, ,, ,, ,, grass ,, ...	357 yds. (326 m, 44)
Minimum distance to unstick—concrete runway	93 yds. (85 m)
,, ,, ,, ,, grass ,, ...	100 yds. (91 m, 44)
Landing distance from 50 ft. (15 m, 24)	294 yds. (268 m, 83)
Landing run using civil operating technique	133 yds. (121 m, 62)
Minimum landing run from touch down at engine on stalling speed	60 yds. (54 m, 86)

MAXIMUM TAKE-OFF WEIGHT AT SEA LEVEL

	lb.
International Standard Atmosphere	13,500 (6123 kg)
A. R. B. Temperate and Arctic Maximum	13,400 (6078 kg)
A. R. B. Intercontinental Maximum	13,000 (5896 kg 56)

The Leonides was subsequently incorporated into the Twin Pioneer and, if the aircraft's single-engined performance caused some concern to potential customers, this was not because the aircraft failed to meet its specification in this respect, but was a result of the limitations imposed on its field performance by airworthiness regulations which had been designed for more conventional airliners.

In this respect the aircraft could perhaps claim to be ahead of its time. Unfortunately, civil operating regulations denied operators the opportunity to take full advantage of the Twin Pioneer's particular STOL characteristics and thus counterbalance the limitations of its cruise speed.

If the Leonides engine presented a problem to the selling of the Twin Pioneer, this was a result not so much of its technical performance as of its nationality. There was a general consensus that, had the aircraft been equipped from the outset with American engines, this would have made a very significant difference to the number of sales achieved, not merely in the United States but in other parts of the world where American equipment and support services were predominant.

This possibility was not ignored by Scottish Aviation. Indeed a number of alternative power-plants were considered over the years including De Havilland Gnomes to be used on a "stretched" version called the Super Twin Pioneer. But when the effects of installing American Pratt and Whitney 600 hp Wasp engines were assessed in the Series 2 Twin Pioneer, the additional weight penalties and poor engine performance unfortunately prejudiced other aspects of the aircraft's behaviour, notably its rate of climb and its operational ceiling.

Nor was it the case that the Company failed to consider other developments designed to enhance the aircraft's suitability for its pioneering role. As well as plans for a supplementary power-plant to replace otherwise essential ballast requirements and enhance performance from high altitude airstrips in countries such as Switzerland, trial installations were pursued to fit the Twin Pioneer with skis and with floats produced on Scottish Aviation's behalf by Saunders Roe.

Method of production was a major part of the project's troubles, since the Twin Pioneer was never properly productionised. Machining work on the early aircraft, for instance, was all done from bar rather than forgings. When the first of the aircraft to use forgings was completed in the Spring of 1959 a marked reduction in manufacturing times was achieved, but in the meantime the earlier method had, like the other development initiatives, simply added to the costs of the project at a time when sales revenues were not being generated as rapidly as had been anticipated. Simultaneously the progressive reduction of production batch sizes in order to limit the Company's immediate financial exposure made it impossible to achieve significant economies of scale.

In April 1958 the Board decided to restrict the Twin Pioneer programme in the meantime to 100 aircraft sets rather than the 200 which it had planned in October 1955. By this time material provisioning was virtually complete up to the 100th aircraft and it is a measure of how slowly sales were achieved, particularly after the Libyan crash, that while orders stood at 61 in July 1958, it was fully five years later before the last Twin Pioneer was delivered to Philippines Airlines in July 1963. Total production was only 87 complete aircraft.

The price of the aircraft was certainly problematic. It was originally envisaged, hopelessly optimistically, that it could be manufactured profitably at a sale price of £25,000. By December 1955 the price had been progressively revised to £42,500 although it was recognised that the margins were still very narrow. Six months later the price had been increased again to a new figure of £48,750 and eventually this price rose to over £60,000. To a large extent, however, this escalation was symptomatic of the project's problems rather than their cause.

These problems did not stem from a lack of interest in the aircraft. Inquiries came from all over the world, including Canada, Australia, New Zealand, Germany, Italy, Portugal, India, Nepal, Ceylon, Abyssinia, Ghana, The Sudan, Libya, Israel, Iran, Iraq, Mexico, Colombo, Honduras, Malaya, Borneo, Dutch New Guinea, The Philippines, Indonesia, Thailand and even Vietnam. In fact, among the lesser known casualties of the war in

In French Guiana, South America

Vietnam and Laos were the five Twin Pioneers which had originally been sold to Philippines Airlines but were subsequently purchased by an operator in Bangkok from where they took part in jungle operations in South-East Asia on behalf of the intelligence agencies.

There were other notable civil sales, either directly or second-hand. The first production aircraft, for example, originally sold to the Rio Tinto Finance and Exploration Company was subsequently purchased by the United Nations in 1962.

G-APHY progressed from service with the Kuwait Oil Company to be used on scheduled passenger and freight services between Portsmouth and the Channel Islands before being purchased by Flight One, used by the British Parachute Association and eventually finding itself, seventeen years after first flight, supporting mining and logging camps in the Northern Yukon. It was a good example of the "Twin Pin's" durability and range of application.

The three aircraft, 513, 515 and 516 purchased by the Iranian Civil Aviation Club in 1957, ended up in Ecuador, Norway and Iceland respectively, while 521, G-APIR, was first chartered to the Iraq Petroleum Company before being used in an attempt to relieve Foula in the Western Isles of Scotland and finding itself storm-bound for 77 days by winter gales in the course of which the crew were presented with an illuminated address by the islanders to celebrate the aircraft's relief effort. It ended its days in Katmandu, a gift from the British Government to the Government of Nepal. 537 came to rest in the same place having started life as G-APMT before being sold to the Red Lion and Sun Organisation, the Iranian equivalent of the Red Cross, and then operating in Sierra Leone.

A number of the final aircraft such as 582, 586 and 590 first bought by the Royal Malaysian Air Force in 1962/63, went on in 1972/73 to Sydney, Australia, where they were employed by Aerial Agriculture.

What is significant about the history of the aircraft's usage is that initial sales by the Company were predominantly to military customers. Of the 87 Twin Pioneers produced, 39 went to the Royal Air Force, and 15 to the Royal Malaysian Air Force. Of the remaining 33, ten were used initially by Scottish Aviation itself before onward sale and only 23 were sold directly to civil customers where the individual orders were obviously small. The range and variety of the second-hand sales of the aircraft are, however, testimony to its suitability to carry out the role for which it was designed.

XM.958 serving with 21 Squadron, R.A.F., in Aden

Even more than with its twin-engined counterpart Prestwick Pioneer sales were heavily biased towards military customers. The Royal Malaysian Air Force bought nine Pioneers, the Royal Ceylon Air Force took four and the only "civil" customers were the Iranian Customs Authority which took two and the Austrian Air Rescue Service which bought one. The most unusual application of any of the Prestwick Pioneers, however, was perhaps when one was chartered by the Twentieth Century Fox Studios to survey French Equatorial Africa for film locations and in particular to search for herds of elephants.

Paradoxically, it was perhaps the aircraft's very specific suitability for its role which was the fundamental factor underlying the commercial failure of the Twin Pioneer project. Many of the countries whose flying needs the aircraft was best suited to meet were precisely those countries which lacked the means or economic infrastructure to purchase it. Certainly difficulties over export credit guarantees and export licences prevented some sales.

Where economic aid was available to those countries it was often applied to the construction of airfields designed to cater for larger aircraft and this detracted from the Twin Pioneer's particular competitive advantage. During his Far East tour in 1955, David McIntyre had noted that the vast amount of money which America was spending on the construction of airstrips in Thailand could be expected to leave little opportunity for British aircraft sales.

His forebodings were borne out. Early in the 1960's negotiations were well advanced for the sale of approximately a dozen Twin Pioneers to Chile and Peru when the US Government intervened by selling both countries a number of second-hand Dakotas for one dollar each, leaving their financial allocations available for the construction of larger airfields.

Prestwick Pioneer cockpit layout

Ironically it was the ready availability of second-hand aircraft, which had provided Scottish Aviation itself with so much profitable conversion work in the aftermath of the war, which now impeded sales of its own product. On one occasion two potential customers from Flinders Island, Australia, arrived at Prestwick to examine the Twin Pioneer, flew the aircraft, declared it ideally suited to their needs and then enquired about the price, on hearing which they announced that in that case they would just go on using their second-hand Ansons, which they had obtained at £15 each!

There were therefore a number of factors which contributed to the Twin Pioneer's failure to achieve the sales which had been expected of it. But if the causes were complex, the outcome was clear. At the end of the 1950s the Company was again haunted by the spectre of financial collapse which it had faced a decade earlier.

Scottish Aviation was never again to manufacture an aircraft entirely of its own design, but this did not mean that the Pioneer or the Twin Pioneer had been unsuccessful designs. On the contrary, both aircraft demonstrated their ability to carry out the job which they were intended to perform.

By the end of the Malayan Emergency in 1960 the RAF's Pioneers had flown a total of 11,132 hours since entering service in October 1953. Their contribution to

"Operation Firedog" involved 37,577 sorties, lifting 29,532 troops and other passengers. Ultimately the RAF took 40 of the total of 58 Prestwick Pioneers which were manufactured between 1952 and 1962. The aircraft served with 267 (subsequently 209) Squadron in Malaya, with 78 Squadron in Aden and with 215 (eventually 230) Squadron in the EOKA troubles in Cyprus. The Pioneer even made unarrested deck landings on HMS Bulwark in 1961 and HMS Albion in 1963.

Prestwick's Pioneers

The last of the RAF's Prestwick Pioneers was not finally struck off charge until, appropriately, New Year's Day 1970. This was XJ465 which had entered service in July 1955.
Given the extent to which the Royal Air Force was the principal customer for both the Pioneers, perhaps the final word on the aircraft should be drawn from that source. In a press release in August 1959 the Officer in Charge of Twin Pioneer pilot training for the RAF stated quite simply,

> "It is an excellent aircraft for the job it is designed to do".

The Prestwick Pioneer and the Twin Pioneer remain the only two aircraft since the 1920s to have been built to designs to have come out of Scotland and the only two ever to have been produced in any quantity. As such they have a unique place, not simply in the affections of those associated with making them, but in the history of aviation.

The shape of things to come

Chapter 7

"DIVERSIFICATION" The Search for Stability

The extent of the crisis precipitated by the failure of the Twin Pioneer may be judged from the fact that as early as March 1956 this project accounted for nearly three-quarters of the factory's workload. The Company had battled hard to win orders in the wake of the Libyan crash, but its mounting financial difficulties finally came to a head on 19 November 1958 when the Finance Corporation for Industry declared that it was not prepared to make any further loans to Scottish Aviation Limited. Coming fully eleven months after David McIntyre's death it could hardly be criticised as a hasty decision or one which had not given the Company the chance to try to recover from the blow which it had suffered with the loss of G-AOEO. The FCI had been extremely supportive and the Corporation's initial commitment in October 1954 to an involvement of £1 million had increased in four years to over £3.5 million.

Among the steps which the FCI took to assist the Company during the period following David McIntyre's death was to introduce the services of a professional consultant from Production Engineering Limited. Bernard Boxall became a regular attendee at Board Meetings from the beginning of 1958, a Director and Deputy Chairman from April of that year and succeeded the Duke of Hamilton as Chairman three years later. The Duke retained a close interest in the Company as its President, but Boxall's appointment reflected the role which he had played in maintaining liaison between the Company and the Finance Corporation for Industry since 1958.

The total capital investment in the Twin Pioneer by the summer of 1958 stood at over £5 million. Scottish Aviation had reported a deficit of £664,000 in 1957 and was already anticipating a larger one for 1958. In the event the loss reached £791,000 and it looked as if the Company would be unable to find the money to pay the wage bill for January 1959 even though the numbers employed had been reduced by as many as seven hundred since the previous August.

In these circumstances it was scarcely surprising that the FCI should conclude that Scottish Aviation could not be expected to continue much longer on its own and that the only hope for its future development would be to follow the Government's prescription for the industry as a whole and negotiate a merger with one of the larger aircraft manufacturing groups in the country. Confidential approaches were made to Vickers, Handley Page, De Havilland, Hawker Siddeley, English Electric and Hunting. Vickers, Handley Page, English Electric and eventually Westland all expressed some interest but the precarious state of Scottish Aviation's finances dissuaded any of these organisations from pursuing a closer relationship.

Ironically it was the Twin Pioneer which now came to the Company's rescue. Arriving back from a sales tour to Portugal, TDM Robertson was met by a grave-faced Bernard Boxall at London Airport and told that a meeting had been arranged for the following morning with Lord Weeks, the Chairman of the Finance Corporation for Industry, because there seemed to be no choice left but to close the Company down. As the Managing Director revealed, however, the Portuguese had expressed an interest in up to 40 Twin Pioneers and the prospect of manufacture under licence in their own country. Armed with this news the Scottish Aviation Directors were at least able to approach the FCI with more than one choice.

Twin Pioneer final assembly line in Fokker Hangar

The Scottish Aviation Board Meeting at the end of January 1959 consequently had three alternatives to consider. The Company could cease operations and go into voluntary liquidation immediately, facing all the implications of broken contracts and the associated liabilities. Alternatively, it could complete its existing contractual obligations for aircraft and spares and close down after that had been done, by approximately the end of April. Finally, if the Portuguese order could be obtained and offered a prospect of profit then the Company could continue production although it was recognised that this would have to be strictly in accordance with orders.

The Directors opted to try and fight their way through. The Finance Corporation for Industry was asked, and agreed, to make the additional investment necessary to pursue this option. Another £100,000 was advanced by way of a loan plus a further £150,000 in exchange for the FCI taking over the assets of the Company's Renfrew operations and of Scottish Airlines.

The prospect of what would have been a very substantial order for Portugal therefore provided a vital breathing space. However, the respite was short-lived. Definition of the Portuguese Air Force's requirements had still not been received by April and was taking much longer to finalise than anyone had originally envisaged. In July it became clear that a decision on the order could not be expected until the following October and The Finance Corporation for Industry concluded that, whatever the technical position, Scottish Aviation Limited was commercially insolvent and must be wound up. It was left to the SAL Board to consult its own advisers and act accordingly.

At this point the Company came closer to closure than at any time in its history. At the Board Meeting each of the Directors in turn expressed the view that there was no feasible alternative. The Managing Director, however, disagreed. As he pointed out, Scottish Aviation had made many friends in the industry over the years and these relationships could now be the Company's salvation, if it could come to an arrangement with its creditors.

The Board was persuaded to give this a try and early in August, by which time all production on the Twin Pioneer had ceased except against existing orders, a meeting was arranged with representatives of the five largest creditors, Armstrong Whitworth Aircraft, Alvis, De Havilland Propellers, Electro-Hydraulics Limited and of course the Finance Corporation for Industry.

The proposal which Scottish Aviation put forward was that the major creditors should agree not to press for immediate settlement of their accounts but allow the Company the opportunity to realise the benefits of the retrenchment which it had undertaken and the good contracts which it still had and so work its way out of its current difficulties. The appeal was more than simply emotional. Complete and immediate closure offered the creditors the prospect of only some four shillings in the Pound in final settlement of their accounts. The alternative of allowing the Company a stay of execution meant that there was a possibility of the creditors securing, if not all, then at least a significantly higher proportion of their money.

Centre sections in the Palace

Prestwick Pioneer G-AOUE was transported by USAF Globemaster for its demonstration tour of the United States, August 1956.

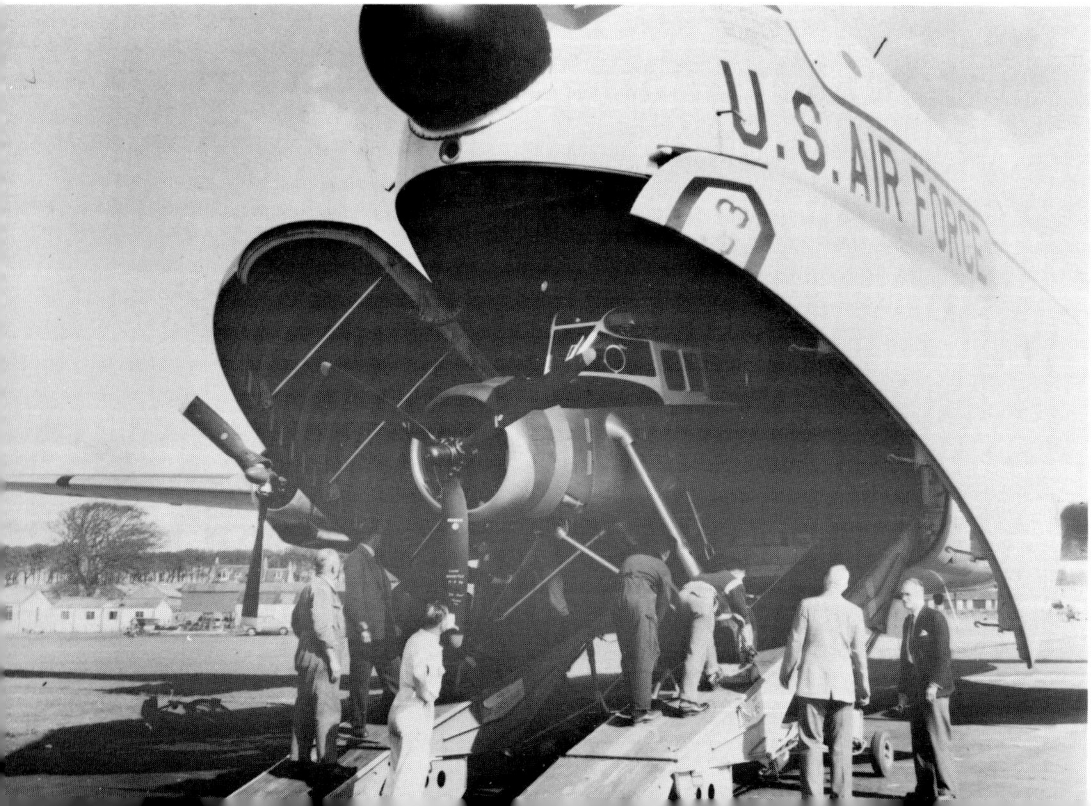

Fortunately the Managing Director and Deputy Chairman were able to convince this meeting to give them the opportunity for which they were looking. The FCI was particularly helpful in stating that, if the Company's proposals were acceptable to the other parties concerned, then it would neither call in its loans nor expect to receive more than a token payment of interest charges each month until the other creditors had been satisfied. It was therefore agreed that Scottish Aviation's situation should be kept strictly confidential and that in the meantime, so as to avoid precipitating the liquidation which all concerned were seeking to avoid, the 700 or so minor creditors should be paid off as quickly as possible. Meetings between the Company and the major creditors were arranged to take place each month to review the position.

Simultaneously the Company sought to mobilise other allies, including political support, in its struggle to survive. Towards the end of 1959 this culminated in a meeting at 10 Downing Street with the Prime Minister, Harold Macmillan, the Minister of Aviation, Duncan Sandys, and John McLay, Secretary of State for Scotland. Scottish Aviation's case was made not simply by its Deputy Chairman and Managing Director, but by Sir Thomas Moore, MP for Ayr Burghs, Mr Willie, later Lord, Ross then MP for Kilmarnock and Mr John Lang, Treasurer of the Scottish Trades Union Congress and Chairman of the Scottish Board for Industry, together with Full Time Officials of the principal Trade Unions represented in the Company.

The object of the meeting was to press the Company's case for the Pegasus, a development of the Prestwick Pioneer into a nine-seater configuration in order to meet an identified requirement for the Army Air Corps. There were those within the Company who believed that this development should have been undertaken earlier as an interim step towards the Twin Pioneer. Whether, if this course of action had been pursued, the Company's history would have been radically different can only be a matter for speculation, but certainly the success of De Havilland Canada's Beaver and subsequently the Otter, with which the Pegasus would have been a direct competitor, revealed a very large market for aircraft of this type.

The Pegasus project never received the Government development contract which would have been required to secure its launch. On the other hand, the Ministry of Aviation did place an order for four additional Twin Pioneers following the Downing Street meeting and this at least provided some alleviation of the Company's difficulties, since it enabled a portion of existing work-in-progress to be completed and sold.

Model of the Mk. IV Pioneer,

The prospect of a Portuguese order for Twin Pioneers had still not entirely disappeared and the Company's agent continued to send progress reports twice weekly until the Portuguese Air Force finally bought an assortment of other aircraft instead in April 1960. On the other hand interest in the Twin Pioneer generally, having recovered from the immediate aftermath of the Libyan disaster, was again being expressed from many parts of the world. Although these possibilities had the value of helping to sustain the creditors' faith in the Company, Twin Pioneer production was never to be anything other than sporadic after 1959.

The Company's recovery plan was based on the principle of "multiplication from division", of diversifying its activities to reduce its dependence on any one contract.

Scottish Aviation's range of interests had always been wide and in the difficult years of reconstruction following the Twin Pioneer project, the contribution made by these ancillary activities became more important than ever before.

Scottish Air Express Limited was a subsidiary which had the distinction of being the first Scottish company to pioneer the development of air freight services. It was originally formed in 1946 to handle incoming and outbound air cargoes at Prestwick Airport on behalf of Scottish Aviation's own airline operation, Scottish Airlines, and also various other international carriers, including KLM, SAS and American Overseas Airlines. Even after Scottish Airlines ceased to perform its own flying operations the name was retained to describe Scottish Express' Aircraft Handling Division whose role was to carry out at Prestwick all the responsibilities which would otherwise have been performed by the airlines themselves, looking after the passengers as well as maintaining and processing the aircraft.

.... *to a major business in its own right*

During the 1960's the impressive list of companies which were prepared to entrust this task to Scottish Express included Aer Lingus, Autair, British Midland, Britannia Airways, British United, Dan-Air Services, Loftleidir Icelandic Airways, Nordair, Overseas National, Pacific Western, Transair, Trans International Airlines, Trans Globe, World Airways Incorporated and the Canadian Armed Forces.

In 1954 the travel agency which had been part of Scottish Airlines was integrated into the freight business and Scottish Air Express quickly consolidated its reputation as a specialist in air transportation. From here it was a short step in 1958 to extend into land and sea transport of freight and passengers as well and early in the 1960's the word 'Air' was dropped from the Company's name.

Pursuing a policy of expansion through acquisition, Scottish Express had, by the time it was eventually separated from the parent Company in 1974, established air freight branches in Prestwick, Glasgow, Edinburgh, Aberdeen, Newcastle and Dundee, trading arrangements at London and Manchester Airports, shipping and forwarding offices at Dundee, Glasgow, Grangemouth and Leith and travel agencies in Ayr, Prestwick Airport, Glasgow, Paisley and Dundee.

Most importantly, in the wake of the Twin Pioneer project at the end of the 1950's, Scottish Express under the direction of David McConnell was a source not only of rapid growth but also of profit and as such was of vital importance to Scottish Aviation.

Increasingly closely associated with the freight and travel agency business from 1958 were two other activities in which Scottish Aviation had been heavily involved since the days when it had been responsible for the Station Contract at Prestwick during the war. The object of integrating the hotel and transport activities more closely with Scottish Express was to provide a comprehensive travel and accommodation service and to be in a position to take advantage of the developments which the Ministry of Transport and Civil Aviation planned for Prestwick Airport in the 1960's.

The extent of the Company's transport activities during the war may be gauged from the fact that on one occasion in October 1944 the American military authorities at Prestwick asked Scottish Aviation to provide 95 drivers to operate its bus and shuttle service within 2 days. After the war the Transport Department under Jimmy Heron ran a daily air freight truck service for agents and airlines long before the latter were in a position to do so for themselves, and this tradition was carried on into the 1970's when the department was responsible for maintaining Scottish Express' commercial fleet. At one time Scottish Aviation itself ran a fleet of Rolls Royce hire cars and more than one employee turned to this source to make the arrangements for a daughter's wedding.

The Company's involvement in hotel management sprang originally from the need to house the trainee pilots of No 12 Elementary Flying Training School. Accommodation was

Rolls Royce for hire

initially in an annexe of the Orangefield Hotel, but later the Company purchased the hotel outright and it developed rapidly under the stimulus of Prestwick's role as the bridgehead for the Transatlantic air ferry.

The Airport Restaurant in the new Terminal, 1964

As a hotel it soon acquired a reputation "the world o'er", and this was due in no small part to the quality of the welcome provided by George Simpson as Manager and Head Waiter Andrew Murdoch, who had a rare gift for making each and every visitor to his dining-room believe that he or she was the sole object of his concern. The list of the distinguished and famous who made a point of using the hotel on their travels and who signed the Visitors' Book is a testimony in itself.

Royalty, politicians, film stars and celebrities of every sort passed through the Airport Hotel and made it another lifeline for the parent company during a dark period.

The efficiency of the Company's catering operation may be discerned from the fact that when Orangefield House was demolished in 1964 to make way for the creation of the new airport terminal and its associated development, the restaurant in the old hotel was closed after lunch had been served on September 18th and was able to re-open in the new building in time to provide dinner the following evening. Scottish Aviation acquired the catering in the new terminal and with it that essential branch of the aviation industry - the duty free whisky concession - but the old magic of the original Airport Hotel had been lost forever together with the fountains amid the palm trees and the control tower perched on its chimney stacks like a stork's nest.

Increasingly the feeding of passengers in the restaurant became subsidiary to the Company's in-flight catering activities and from 1st December 1974 Scottish Aviation's Catering Division became part of Scottish Express.

The Company's involvement in civil engineering also went back to its earliest years, to the establishment of the airfield at Prestwick and the rapid development of facilities required in the context of war. Among the projects for which Jack Dickie's Contracting Department had been principally responsible were the creation of Grangemouth Airport and the movement of "The Palace" from Bellahouston Park in Glasgow and its re-erection at Prestwick. Both tasks were fraught with difficulties. In levelling the ground at Grangemouth during the Summer of 1939 the Company was restricted by an Air Ministry stipulation that no more than one acre of turf could be lifted at any one time as a precaution which would allow the existing surface to be re-laid within two days to make the airfield operational for the defence of the Firth of Forth. Nevertheless the construction of the airport was completed in record time, using various examples of native ingenuity including the use of a mole plough to lay underground cables, an idea which was observed and copied by a neighbouring contingent of Royal Engineers who went on to employ the technique very successfully during the invasion of Europe.

The approach of the Contracting Department perhaps above all others in Scottish Aviation was one which put results first and bureaucracy second. When the Palace had to be camouflaged in 1940, the Camouflage Officer sent to advise the Company was a young man who had, as a civilian, been an artist at Elstree Film Studios. His duties required him to inspect the results of the camouflage from a height of 12,000 feet, but when Noel Capper promptly offered to oblige with a flight in one of the Company's Ansons, the young man insisted that, in line with the Safety Regulations, he must be flown by a regular RAF Officer. With some difficulty this was arranged. What the Camouflage Officer could not conceal, however, during a bumpy descent was his own air-sickness and he had reason to rue his earlier insistence when, on landing the Anson, his RAF pilot sat back in his seat and announced, "Not bad for my first time in a twin-engined aircraft!".

Two views of the Palace of Engineering in Glasgow before it was dismantled and moved to Prestwick

The jobs undertaken by the Contracting Department in later years were also many and varied, ranging from the building of houses in Symington to the erection of the "Measured Mile" markers on the Island of Arran for ships undergoing their sea trials in the Firth of Clyde. Regular work came from maintenance contracts at various military establishments in Scotland and the North of England including Wig Bay, West Freugh and Carlisle, and there were occasional contracts for the Government too, notably the construction of a series of radar stations during the later 1960's on behalf of the Board of Trade.

While the Company's contracting business did not take it "the world o'er" in the way that some of its other activities did, it nevertheless found itself in some of the remoter parts of Scotland at one time or another. Probably the contract with the most complicated logistics was for the missile tracking station on the Island of St Kilda in the Outer Hebrides 140 miles west of the Scottish mainland. When, in 1971, Scottish Aviation won the contract to re-surface the access road and slipway, reinforce a Bailey bridge and repair the station's buildings, the island's only other inhabitants were a few wild sheep and legions of sea birds, including approximately one-third of the World's entire population of gannets.

The job had to be completed between May and September, because the high seas and violent winds around the island meant that the only landing place was a sandy beach which was formed by the Spring tides and washed away again every Autumn. Personnel, equipment and supplies had to be transported by landing craft in a journey which took the best part of two days and was not for the faint-hearted. On one occasion the SAL party was caught in mid-ocean by a strong gale and had to be escorted back to the mainland by a Royal Navy destroyer.

Like aviation, building contracting had its peaks and troughs, but if it could not claim to have been a consistently profitable branch of Scottish Aviation's activities, it was nevertheless another means by which the effects of the acute downturns in the aircraft business were mitigated.

An old friend in adversity which made a timely re-appearance in the Company's affairs at this time was was the Dakota. In 1961 and 1962, following work done for the Irish airline Aer Lingus, conversions were again being requested from as far afield as India, Argentina, Canada and the Sudan.

Another aircraft from the Company's past which suddenly figured again in 1962 was the Skyraider and in this case it was twelve of the very aircraft which the Company had originally modified for the Royal Navy in 1954 which now returned to Prestwick to be converted into target tugs for the Swedish Air Force.

Handover of first converted Royal Navy Skyraider to Swedish Air Force, September 1962

Although the prospect of a further 18 Skyraider conversions for Brazil came to naught because of the uncertainty as to whether the Company would be paid, these conversion programmes pointed the way to a new venture which was to become a sizeable element in Scottish Aviation's diversification plan.

SALchek was formed in June 1962 to offer an aircraft overhaul, repair and maintenance service which would capitalise on the experience which the Company had acquired in handling a wide diversity of types over the years. In its first month of operation, twelve aircraft passed through its hands, comprising six distinct types ranging from the 3-seat Auster Alpha to the 120-seat Lockheed Starliner.

By 1964 nearly 30 operators were using SALchek's aircraft maintenance service and in that year the Service Department was called out 243 times, its Engineers travelling "The World O'er", from Norway to Malaya and from Bahrain to the Bahamas. The following year they were representing Scottish Aviation in another air lift operation to add to the Company's previous ventures in India and Berlin, this time in support of the Scandinavian Flughjalp relief to war-torn Biafra.

It soon became apparent that the Company needed even larger facilities to be able to work on some of the aircraft which SALchek was being asked to handle. In 1965 a new hangar with a floor area of 45,000 square feet and a concrete floor 12" thick to take heavy wheel loads was constructed to house the largest aircraft then in civil operation. The "Britannia" was the most spacious hangar in Scotland at that time and very soon it was housing not simply the aircraft which gave it its name but also a major programme of Viscount conversions, chiefly for Aer Lingus.

Typically there was still the occasional visitor that was difficult to accommodate, but the SALchek team was not easily put off and when a United States Air Force Globemaster with a wing span 23 feet wider than the hangar arrived requiring a damaged rudder hinge bracket to be repaired, its massive 48 foot high fin was squeezed under cover by letting down the main undercarriage struts and tyres.

Laker Airways Britannia in the "Brit" Hangar, 1966

The first of the Viscounts for conversion, 1963

Freight door installation on Aer Lingus Viscount, 1966-67

In 1968 the Service Department handled a total of 566 aircraft, principally Douglas DC8's and Boeing 707's but ranging down to types as small as the Tripacer with a wing span of less than 30 feet and a weight of only 2000 lbs.

"SAL cheking"

With a view to expanding the SALchek operation, consideration was given early in 1966 to establishing a satellite operation at Glasgow Airport and to securing designation as an official Repair Station for the British Aircraft Corporation. Vickers, however, required a 10% commission on all work related to BAC aircraft in exchange for this privilege to which Scottish Aviation's prompt response was to drop the whole idea.

Profit was always difficult to make on SALchek work for two reasons. First, it was almost impossible to give accurate quotations to customers because, until the aircraft had been received and surveyed, the full extent of the work required could not be established. Secondly, as the sheer variety of aircraft handled by the department suggests, there was considerable difficulty in obtaining continuity of work on the same types.

The problem of continuity, although never entirely absent, was far less acute in two other areas of overhaul which were consequently to become mainstays of Scottish Aviation's business during the years of diversification following the Twin Pioneer.

The Company had sought to build on its war-time experience of engine overhaul and servicing as early as December 1945 when it became an official distributor and repair centre for Pratt and Whitney. Unfortunately this work did not develop as extensively as had been hoped and an early contract from Rolls Royce to overhaul Packard Merlins built in America during the war also turned out to be disappointingly short-lived, lasting only from 1952 to 1954.

S.A.E.S. at work on Pratt and Whitney engines

In May 1959 the Engine Shop was just breaking even when a request was received from America for the Company to quote for the overhaul of 78 Pratt and Whitney R2000-4 engines of the sort used in the DC4 Skymaster. Coupled with work already being carried out on the R1830, the DC.3 powerplant, this offered the prospect of an expanding workload from 1960 to another SAL subsidiary, Scottish Air Engine Services.

SAES was originally a collaborative venture between Scottish Aviation and Air Engine Services of Lingfield, in which the English firm was to be the sales company and Scottish was to provide the actual engine overhaul facility. In practice the collaboration was not entirely satisfactory and SAES became a wholly-owned subsidiary of Scottish Aviation from 27 December 1963. By this time, after a slow start, R2000 engines were arriving in significant numbers and the throughput of Pratt and Whitney engine overhauls in the "Tiger" Hangar, which had originally housed the aircraft of No 12 Elementary Flying Training School, had increased to ten per month.

Pratt and Whitney work was, however, rapidly eclipsed in scale by contracts from Rolls Royce with whom Scottish Aviation's earlier association was renewed in 1961 when the Company undertook the manufacture and overhaul of cylinder blocks for the Merlin, the engine which had powered so many famous aircraft, including the Hurricane, Spitfire, Lancaster and Mosquito. The significance of this contract for the recovery of Scottish Aviation's fortunes is readily apparent from the fact that the work was valued at £500,000 per annum and was expected to last until 1965. In the event the last Merlin to be overhauled on a production line basis in Great Britain was completed at Prestwick in April 1967, although - like the proverbial old soldier - the engine seemed to be immortal in a literal as well as a metaphorical sense and the huge unmistakable packing crates have continued to appear from time to time since then.

Like other contracts before and since, it seemed for a while as if the Rolls Royce work was not going to fulfil the expectations originally held of it. In August 1962, after more than 80 Scottish Aviation personnel had undergone a lengthy period of training at Hillington, the workload suddenly went flat and thirty men had to be paid off, although Rolls Royce took the unusual step of keeping some of the Scottish Aviation men in its own employment until such time as the work again began to flow into Prestwick.

After this worrying start the supply of work from Rolls Royce went from strength to strength. The Merlin cylinder block contract was soon extended to include the complete overhaul of the engine and was followed in 1965 by work on Griffons, Derwents and Sea Griffons, the powerplant used in the RAF's high speed rescue launches. By the beginning of 1963 Scottish Aviation had so clearly established itself as something more than the normal sub-contractor in its relationship with Rolls Royce that, when Rolls' own factories went on to short-time working, no reduction was made in the workload at Prestwick.

By 1966 Scottish Aviation had taken over complete responsibility for Rolls Royce piston engine overhaul and spares production as the factories at Hillington and East

The last Merlin

Kilbride completed their own transition on to gas turbine work.

Like all major new projects the Rolls Royce association was the occasion for a further development of the facilities at Prestwick. In this case a Heat Treatment Shop was created to provide for the carbonising, hardening and tempering of ferrous parts like cylinder liners and camshafts, while the Wet and Thermal Treatments Department was re-equipped to make it capable of carrying out an expanded range of anodising and plating processes.

Over the years the Company worked on a wide range of Rolls Royce engines including overhaul of the Dart, the earliest of its gas turbines, and of the Meteor, a development of the Merlin which was used to power the Army's Centurion tanks. Parts were produced for the Spey, Dart, Tyne, Avon and Conway engines.

The Griffon, production line

But it was the Griffon which was the backbone of the work and by the time that the programme finally came to an end at Prestwick in 1981 as a result of the Ministry of Defence decision to phase the RAF Shackleton out of service in favour of more modern types of airborne early warning aircraft, the factory, as Scottish Aviation and as British Aerospace Scottish Division, had overhauled a grand total of 1,910 of these engines.

The association with Rolls Royce was also one of the means by which the Company was able to employ and maintain its design capacity as the Twin Pioneer project came to an end. By April 1961 the Design Office had a complement of 16 engaged purely on work for Rolls and when in 1964

Scottish sought sister design rights for the American Phantom aircraft which was to be acquired by the Royal Navy, it was eventually proposed that this should be done on the basis of a joint submission from SAL and the Rolls Royce Scottish factories. Other sub-contract design work in this period was undertaken for Short Brothers and Harland of Belfast on the Skyvan I and also for Ferranti.

The other association which was not new but which acquired a vital significance for the Company in the 1960's was with the Royal Canadian Air Force. Originally based at Renfrew, where it had overhauled Sabres for the United States Air Force for 18 months from November 1954, SAL had been involved since 1955 in the servicing of a series of RCAF aircraft, including Sabres, CF100's and T33's.

By the Spring of 1959 the scale of this work was approaching one million manhours per annum, but the Renfrew factory was being criticised by the Canadian authorities for failing to meet its programmes. In fact the Canadians were on the verge of transferring the contract to Bristol when SAL proposed that the work be moved to Prestwick in view of the continuing growth which was anticipated and the better facilities which existed there for this type of work.Apart from any other consideration the Company had received a constant stream of complaints from Renfrew about the noise of the jet engines. The inhabitants of Prestwick and district, who had once been affronted by the "din" of the Company's Tiger Moths, had long since grown accustomed to the sounds of the airport.

Despite the complexity of the employee relations involved and the sheer logistics of maintaining programme, the transition took place smoothly during the summer of 1960. A large part of the SAL site at Renfrew was taken over by British European Airways.

At Prestwick the prospect of this sudden influx of work, coupled with the impending demolition of the Company's earliest hangars as part of the British Airport Authority's development of the airport, meant that despite the run-down of the Twin Pioneer programme, there was likely to be a shortage of hangar space to accommodate the RCAF aircraft. Plans were laid to build a "Bridge Hangar" spanning the gap between the existing No 11 and No 12 hangar complexes, and since the Company's own finances clearly could not stretch to an investment of this magnitude, the Government was pressed to provide the funds, partially by way of compensation in kind for the buildings which the Company was about to lose. However, Duncan Sandys, the Minister of Aviation, although he subsequently visited Prestwick at the instigation of the

Secretary of State for Scotland and was impressed with the work being done there, was not prepared to approve the construction of the hangar. His view was that the process of amalgamation, which the Government was forcing on the industry, meant that there was no long-term future for the aircraft industry in Scotland.

As it transpired, the absence of the Bridge Hangar became unimportant because by the end of 1960 the Canadian authorities had decided to phase the Sabre out of service in Europe. One potential embarrassment was replaced by another. SAL was now confronted with the prospect of a six month period between March and October 1961 when its RCAF workload could be expected to halve.

Sabre, CF.100 and CF.104 of the Royal Canadian Air Force supported in Europe for 23 years by Scottish Aviation

The hope was that thereafter the new high performance strike reconnaissance aircraft, the CF104 would be allocated to Scottish Aviation and would more than fill the gap.

Yet it was by no means a foregone conclusion that the 104's would come to Prestwick. A satisfactory record of past performance was not sufficient in itself. As with all military aircraft projects there were also political considerations. In December 1959, when SAL was pressing the Government to take its Pioneer development, the Pegasus, to satisfy the Army's requirement for a liaison aircraft, the Canadians had not hesitated to point out that they would be "most upset" if their Beaver lost this order, valued at £2,500,000, in view of the fact that Scottish Aviation's RCAF work was worth over £1,000,000 per annum.

CF.100 and CF.104 overhaul lines at Renfrew and Prestwick.

SAL cocooned the R.C.A.F. Sabres as they were phased out of service

The Company considered it particularly vital to build on its previously successful relationship with the RCAF and thereby secure the CF104 maintenance and overhaul work, because the German, Danish, Norwegian, Belgian and Dutch Air Forces were also scheduled to equip with the aircraft and Scottish Aviation was anxious to make what it could of these opportunities.

The German work in particular was pursued with vigour and after a visit by the German Air Attache in November 1961 the Germans went as far as to draw up a contract for the overhaul of 75 Sabres. Negotiations with the German authorities for 104 work went on for another year but lapsed after it became apparent that it would require an additional capital investment of up to £100,000 which the Company simply did not have. Whether the tragic accident record of the Starfighter in German service would have been any different had the overhaul and maintenance of the aircraft actually come to Scottish Aviation can only be a matter for conjecture, but certainly many Canadian pilots and their families had cause to be grateful that the Canadian Air Force's CF104's did not fail in the same way as the German.

SAL finally secured the Canadian 104 contract in the summer of 1962. Being a significantly more sophisticated aircraft than its predecessors the Canadian authorities were adamant that it needed a higher calibre of maintenance engineer and had insisted that a number of Scottish Aviation employees should undergo a formal examination to establish whether they were suitable. All passed with flying colours and a party of 29 went to Canada in the middle of August to receive training on the aircraft.

The first CF104 was flown into Prestwick from the Canadair plant in Montreal on 15 January 1963. Two years later the programme had grown to a rate of 14 aircraft a month, rising to 18 from April 1965. Working with a characteristically Scottish disregard for other people's holidays, the 1,000th CF104 test flight from Prestwick was squeezed in on Boxing Day 1967 before the programme was five years old.

Another memorable date came on 18 January 1970 when aircraft 104751, after complete dismantling, structural sampling inspection and re-assembly, was test flown without a single defect and was declared by its Canadian pilot from the Central Experimental and Proving Establishment detachment at Prestwick to be the most snag-free aircraft he had ever flown. The evidence that this was typical of the quality of the Scottish Aviation workmanship on the programme lies in the fact that the average number of test flights required was only some one-and-a-half flights after a run of 600 aircraft.

The RCAF's CF104's in Europe were based in France and Germany at Marville, Soellingen and Zweibrucken. Throughout the project SAL sent Mobile Repair Parties abroad to service the aircraft at the Wings and initially all the work was done there until the Company succeeded in persuading the Ministry of Aviation to allow arrestor gear to be installed on the runway at Prestwick in the summer of 1964.

When the 104 programme was first envisaged, it was expected to provide work for five years. In the event it lasted for fifteen before it was overtaken by the ever-present political pressures and the work was taken back to Canada itself to offset the effects of the recession in its own aircraft industry.

Scottish Aviation's involvement with the RCAF extended for almost 25 years in all. The importance of this continuity to the Company during a period which was often fraught with other difficulties is difficult to overestimate. The length of the association is also particularly striking in view of the considerable scepticism originally expressed by the Canadian authorities regarding Scottish Aviation's ability to carry out the work.

Overcoming this initial opposition owed much to the quality of the work produced but a great deal also to the personal relationships which were established between the Company and the Canadians and the high regard which the Canadians developed for particular individuals, such as Jimmy Mitchell, then Works Manager at Prestwick and, most importantly perhaps, "Bill" Whyte, who was persuaded to become Scottish Aviation's representative in Canada in

1957 and who became the lynch-pin of the Company's relationships with Canadair, the RCAF, Avro Canada and the Department of Defence Production.

There might perhaps have been some particular affinity between Canada and the old country from which so many of its settlers had originally come. Certainly both shared a passion for "the roaring game", and these invisible bonds became apparent when SAL instituted a trophy for a curling competition on the RCAF base at Rockcliffe in 1958. On another occasion, at a dinner given by the Company for a number of VIPs from the Canadian Department of Defence Production, the guests absorbed the spirit of the hospitality so thoroughly that the Director of the DDP insisted on all the gentlemen present placing one foot on the table in order to drink a Highland Toast in the traditional way, while the Airport Hotel's Head Waiter swiftly summoned his staff to take up positions from which they could catch the glasses as they were thrown over the shoulder.

A strong personal rapport characterised the other major sub-contract which began in this period and, like the Rolls Royce and RCAF contracts, lasted throughout the remainder of the Company's life as Scottish Aviation Limited and into the British Aerospace era.

It had perhaps been an omen of things to come when the first CF104 to arrive at Prestwick had been a passenger in another aircraft designed by the Lockheed Aircraft Corporation of the United States, namely the C130 transport, commonly known as the Hercules.

The Lockheed C.130 "Hercules"

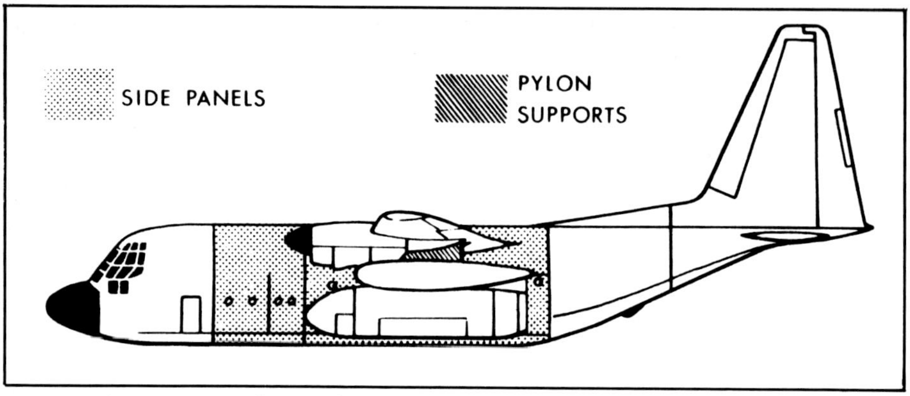

Hercules structure manufactured by SAL

On 2 December 1965 Scottish Aviation secured a contract from Lockheed to manufacture side and top fuselage panels, forward fuselage sections and landing gear panels, comprising in all some 40 feet of the main fuselage plus the wing tank pylons for the 48 Hercules aircraft which had then been ordered by the RAF. Although the manufacturing work represented part of the offset arrangement negotiated when the RAF deal had been concluded with the Americans, SAL did not receive the work as a matter of course. It was won in the face of stiff competition from Hawker Siddeley, The British Aircraft Corporation and Short Brothers of Belfast and only after the Company had satisfied Lockheed-Georgia that it was capable of performing to the exacting standards which that corporation demanded of its sub-contractors.

After meetings with Lockheed representatives in London, a Bidders' Conference at Marietta and a visit by the Americans to Prestwick, the SAL team of four which was spear-heading the negotiations found itself back in Georgia. After coping with a third degree interrogation from a panel of nearly twenty Lockheed personnel the Scots felt they must have done enough to make their case and said so only to be told that they still had to meet "Big Daddy".

Big Daddy was Lee Poore, Lockheed-Georgia's Director of Manufacturing Operations whose first reaction on hearing that a bid had been received from Scottish Aviation was simply "Who the hell are they?" When he joined the meeting to scrutinise the Scottish Aviation team, he left them in no doubt as to what criterion he would use in assessing their suitability. "Skedule", he said, "is our god". SAL's Managing Director looked him in the eye, replied simply, "We worship at the same shrine", and knew from the look which shot back across the table that the contract had been won.

Certainly a few prayers were offered up between then and the following August when the first deliveries were due to be made, notably when a crate which was supposed to be carrying urgently needed drawings from Georgia turned out to contain oil paintings.

Nevertheless, during the factory's summer "holiday" in 1966 two lorries carrying 3,700 cubic feet of crated assemblies pulled away from Prestwick to Grangemouth Docks en route and on schedule.

The hangar which had once housed the Fokker "flying classroom" airliners now became "Hercules Country" and again the quality of the output quickly became more than just its own reward. By the beginning of 1967 the original order had been extended to 101 sets and, as the years passed, this went on increasing and was supplemented in 1973 by work for Lockheed-California on the manufacture of cargo doors, escape hatches and avionics doors for the Tristar.

The 100th set of C130 panels left Prestwick on 3 May 1968. The 200th followed it just over two years later in June 1970 and the 300th in February 1973. From the 357th set transportation to Georgia was by means of high cube containers from Greenock via Charleston in South Carolina. When loaded there was less than three-quarters of an inch clearance between the end of the transport cradles and the container doors but the method provided a "door-to-door" service to the production line. When a seamen's strike in the USA threatened to disrupt the flow in 1969, two sets were even loaded into pallets specially designed by SAL and collected from Prestwick by Lockheed's stretched C130 demonstrator, the C100-20.

The supreme accolade to Scottish Aviation's contribution to the most successful military transport aircraft in the world came on 2 June 1969 when Lockheed itself presented the Company with its exclusive High Standard of Quality Award, a commendation which had at that time only been won by seventy out of a total of over 8000 suppliers and sub-contractors. Scottish Aviation, the 68th company to receive the award, had the unique distinction of being the first company outside North America to do so.

From the beginning of its involvement at No 4202 until the end at No 4976, Scottish Aviation manufactured a total of 773 ship sets of C130 fuselage panels and pylons over a period of seventeen years. As with the Dakota the Company had formed a special association with an enormously successful aircraft. It had worked hard to gain a share in that success and could take pride in the fact that it had in its turn contributed to it.

"Hercules Country", the Fokker Hangar

From the Company's perspective the Lockheed contract was another vital element in the programme of diversification which spread its activities over a portfolio of major projects and brought it first a measure of stability and then of prosperity in the years following the disappointment of the Twin Pioneer project.

In November 1960 Lord Weeks, Chairman of the Finance Corporation for Industry and a great friend to Scottish Aviation had died. Fortunately he was not the Company's only proponent at the FCI. Among the others the one who was perhaps most closely involved with SAL's affairs, and who eventually served as a Director from 1964 until it became part of British Aerospace, was G S Nelson.

In June 1961 it was agreed that the Company's debts to its major trade creditors would be discharged on the basis of payments such that a total of 25% of the original indebtedness would have been settled in cash plus the issue of unsecured loan notes, free of interest, amounting to a further 29%. In the event of the loan notes not having been repaid within seven years, then the creditors were left with the right to demand repayment on or after 30 September 1968. In the meantime the Company was prohibited from declaring or paying any dividends on its share capital and, more importantly perhaps from the point of view of its capacity to recover, from making any capital investment in excess of £50,000 in any financial year. The FCI for its part generously accepted repayment of a little over 2% in cash and some 9% in loan notes and proposed that the other major creditors should be repaid in full before itself.

In the event the outstanding notes were paid off well within the term of this agreement and the Company was able to introduce certain expenditures which it could not have contemplated while it was fighting for its survival. Awards to employees who had achieved 25 years or more service with the Company were introduced early in 1964. The celebrations originally took the form of afternoon tea rather than an evening dinner but nevertheless the ceremony was indicative of a new optimism. The same impulse was evident in the introduction of a Works Pension Scheme in the same year.

The first long service gold watch presentation, 1964

The Company was surrounded by change. The Tiger Hangar, the Anson Hangar and the Orangefield Hotel, cornerstones of its history, were all scheduled for demolition. But the Company itself was looking not to the past but to the future.

"Every home should have one ...."

Chapter 8

"UNCO VARIORUM" Other Projects

By the time that Scottish Aviation celebrated its Silver Jubilee in 1960 it could fairly claim that, despite some turbulent fortunes en route, it had succeeded in realising its founders' objectives. At one time or another, although not always for as long as it would have wished, it had established in Scotland all the major branches of the aviation industry. In the process it had also sprouted a few branches which had little or nothing to do with aviation. What made Scottish Aviation unique in the industry was the sheer diversity of work which it undertook in the course of its history.

To some extent this was a matter of necessity rather than of preference. At critical periods in its history Scottish Aviation's principal concern was the business of survival and many of the projects which it undertook were expedients designed to fill its capacity, to preserve the skill resources which it had built up and to exist until another mainstream opportunity could be secured.

This had been the position for the factory during the recession in the industry after the war and it was to be so again in the wake of the Twin Pioneer project. Among the more outlandish projects on which the Company embarked at this time were the production of spectacle cases, of dies for manufacturing fountain pens and of machines for activities as diverse as perforating paper, packing cigarettes and peeling oranges and lemons.

Not all of the fill-in projects were as far removed from the field of aviation as these. One of the Company's lesser known ventures which began in 1965 and was still going twenty years later was the design and manufacture of hydraulic seat locks to control the recline angle on aircraft seats. Marketed by McEwan King Components the embodiment of these parts meant that Scottish Aviation was able to make its own small contribution to aircraft such as the BAC One-Eleven, the HS.125 and even Concorde.

The Scottish Aviation Orange and Lemon peeling machine

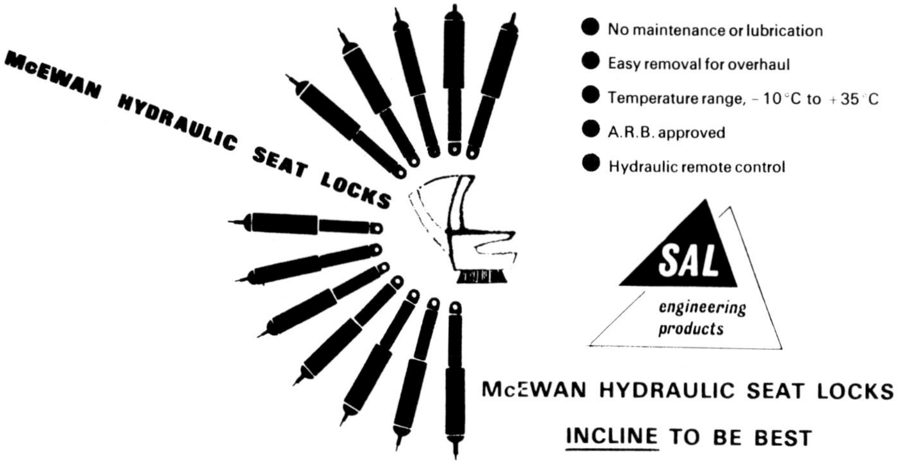

The policy of diversification on which the Company embarked in the 1960's, however, was based on a recognition that it was not possible to fill the Prestwick factory with aircraft work and that a diversion into commercial engineering activity was essential. In 1963 therefore the Company was consciously re-organised into two principal operating divisions, one dealing with aircraft manufacture and maintenance while the other handled general engineering. The objective was to achieve a balance in the workload which would offset the Company's vulnerability in the event of reductions in its two major contracts for Rolls Royce and for the Royal Canadian Air Force.

In addition to seeking new general engineering work from external sources the Company looked to achieve this improved balance through the creation of a Project Department designed to conceive and evaluate its own projects. Established in May 1963 under Dr Gordon Watson, who eventually succeeded Robert McIntyre as Chief Engineer and went on to become a Director in 1969, part of the Project Department's remit, implicitly if not explicitly, was to extend the diversification process away from the aircraft industry. As a result the Company again found itself engaged in a fascinating variety of products.

An early brainchild adopted by the new department was the Artplax panel. This was a lipped metal panel, produced on the rubber die press, backed by fibreboard and finished either in plain high gloss stove enamel or by Silk Screen printing in decorative design. It was used for wall cladding, partitions, advertising hoardings and murals in locations such as shops, offices and hotels. The panel was held in place by means of a dimple which fitted into a spring tee section.

From aircraft design to interior design : Artplax panels

A subsequent development added small pressure jacks to enable a multiple-panel unit to be self-supporting without the need for nails or screws. At a time when pre-fabrication was gathering pace in the building industry, Artplax seemed to be an extremely promising project, which would utilise otherwise idle capacity on an item which had the merits of being simple to produce and offered a substantial profit margin.

Selling at between £3.75 and £4.00 per square yard in 1963 the panels compared competitively with tiled or mosaic walls which were some 25% more expensive. SAL was quickly approached by an agent eager to market the panels through a chain of nearly two-hundred outlets throughout the UK and Ireland. A sizeable number of enquiries followed from America and Artplax panels soon stretched from the Scottish Express Counter in Prestwick Airport's Booking Hall to the Olympic Exhibition Hall in London via the name boards on the platforms of Birmingham's New Street Station.

Less successful was the product christened SALrad 75. This was a panel heater which utilised the properties of a conducting rubber only recently discovered by I.C.I. The rubber provided a thin but flexible heating element in a casing formed principally from sheet metal and again provided work of the sort at which the Company was adept from its aircraft activities.

The SALrad 75 radiant panel heater

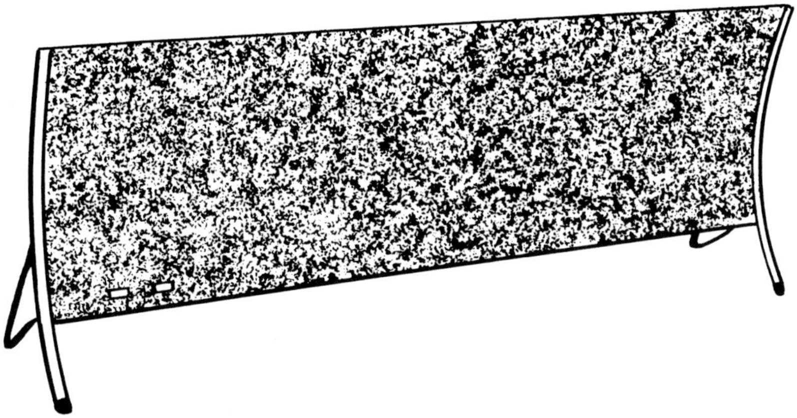

The conducting rubber element was not capable of running at a very high temperature without disintegrating so SALrad 75 with its output of only three-quarters of one kilowatt was designed to be a radiant panel heater which would be used for background heat and which would be not only more attractive than the bulky oil-filled alternatives but also cheaper to run. In July 1964 the go-ahead was given for the manufacture of five hundred units, but a year later the Company decided to discontinue production and to sell off existing stocks. The product itself was regarded as satisfactory but not one which SAL could profitably handle without experience of the electrical goods market. Steps were therefore taken to sell the idea and patents to a company which was in a position to market the radiator.

Perhaps it was not altogether surprising that the Company had tried to become involved with heaters since it had already had some success in making coolers. In 1960 negotiations with the Heat Pump Refrigeration Company of London had led to the establishment of another of the collaborative ventures which characterised much of Scottish Aviation's diversification effort. SAL's Irvine factory, which had originally been a soap works and in which soap still oozed through the floor and hung in stalactites from the ceilings when it had been taken over for the manufacture of Twin Pioneer components in the 1950's, consequently changed its role again. Under the name of Heat Pump and Refrigeration (Scottish) Limited it now began to manufacture cooling units to designs supplied by the English firm.

Bottle cooling tray manufactured in the Irvine factory

The units in question were principally bottle cooling shelves and cellar cooling equipment. The bottle cooler was a flat metal tray with a wooden surround and a refrigeration unit mounted either on or under the tray or remotely under the bar counter itself or even in the cellar. The most critical parts of the construction were the brazed joints of the piping carrying the refrigerant circulated by the compressor. The electronic leak-detector attached to the unit was designed to be so sensitive as to indicate a seepage of Freon gas as minute as half an ounce per annum.

Despite extensive interest in the product, the achievement of sales to prestigious organisations like Guinness, which took a hundred units, and exports as far afield as Canada, Denmark, Malta and Austria, the project never developed on the scale that had been anticipated and in 1964 the production line at Irvine was closed and the residual work transferred to the Prestwick factory.

Other projects from this period of a rather different nature included two with medical connections. One was the manufacture of the Strontium 90 irradiation unit for the treatment of skin cancer by the Radio Therapy Department of Glasgow Western Infirmary. The machine was both designed and manufactured by Scottish Aviation to a specification provided by the Regional Physics Department of the Hospitals Board. The unit employed a low power radioactive source in the form of a 4" x ½" plate supplied by the UK Atomic Energy Authority and located inside a barrel-shaped head lined with 800 lb of lead for protection. The head was mounted in such a way that it could be turned in any direction, raised, lowered and tracked along beams fixed into the walls of the operating theatre.

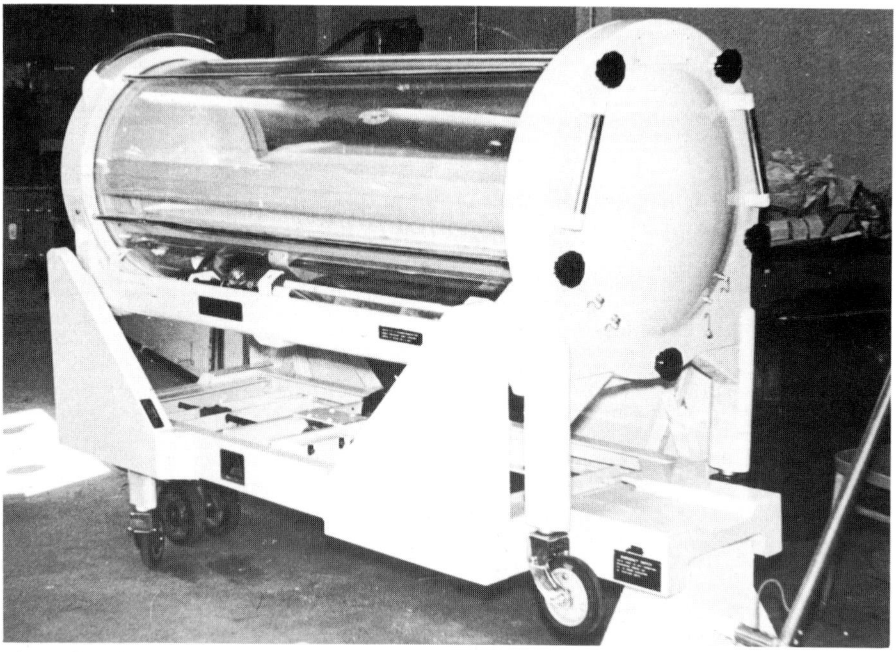

As well as hospital beds SAL manufactured this hospital trolley

The other was the production of a batch of hospital beds for the Ministry of Health. Design was by the Royal College of Arts and the bed was intended to be a universal, "multi-role" model utilising a variety of hydraulic devices to make it capable of equipping all types of hospital. The factory's fitting, welding and paint shops were all heavily involved in the project and, like the light alloy bus bodies before them, the beds had the benefit of the Technical Office's stress calculations. If anything the beds were treated even more like aircraft than the buses and had their own resident Scottish Aviation Service engineer for four weeks when they were first installed and tried out at Chase Farm Hospital in Sussex.

These projects were many and varied but they lacked the promise of continuity which the Company was seeking. In a classic example of history repeating itself the Project Department therefore came to the conclusion that the best opportunities for new ventures lay in the field of motor transport, although the Board had already rejected the idea of machining motor car components because of the poor industrial relations record of the car industry and its disruptive effect on its suppliers.

After the losses which had ultimately been incurred on its bus bodies in the late 1940's and early 1950's this might have seemed an alarming conclusion, but on the other hand the tractor cabs which SAL had designed and built since 1948 were still in production in 1960, although - or perhaps because - the range had by then been reduced to only two types.

When the Company embarked on a new programme of vehicle body-building in 1963 the Board was clear that the production must use commercial rather than aircraft criteria. The result was the manufacture of a series of bodies some of which, including travelling shops, were designed by SAL itself and some of which sprang from an association with Bonallack Commercial Vehicles. The first of the vehicles bearing the "SALeng" label appeared at the Scottish Motor Show in November 1963 and by the time that the vehicle body section was finally closed in 1969 Scottish Aviation had built and sold over 450 commercial vehicles.

The most novel of the vehicle body projects was the Minatic semi-trailer. This was designed to provide an economic answer to the need for road transport of lightweight, high volume goods. Applying aircraft manufacturing methods and materials the Company produced a series of fully-articulated semi-trailers to be used in conjunction with small prime-mover pick-up vehicles from the Austin, Morris, Ford, Bedford and Commer ranges.

The numeric designation of the various Minatic models referred to their trailer load, the Minatic 7 taking seven cwt, the 11 eleven cwt and so on through 21 and 31. In terms of carrying capacity these trailers provided 240, 300, 500 and 576 cubic feet respectively. Rapid turn-round was another prime feature. The towing fixture

Commercial vehicle bodies made by Scottish Aviation 1963-1969

was a positive lock hitch connected to a specially designed "spider" secured by four bolts in existing sockets on the prime mover sidewalls. As a result it took less than a minute to convert either from truck to articulated unit or back again. Furthermore, no modification was required to the prime mover beyond the addition of a socket and switch to connect the rear, stop and indicator lights to the trailer. The Minatic therefore offered the advantages of low initial cost and low operating costs. The polished aluminium alloy construction avoided the problem of rust and meant modest maintenance costs, while the Minatic 7 only increased the road tax payable on the prime mover by £3 per annum in 1964. In terms of fuel economy the performance was also impressive, a fully laden Minatic 7 with an Austin 5 cwt pick-up returning a figure of 37.5 miles per gallon in an independent road test.

The Minatic Semi-trailer

Although the trailers were designed and manufactured by Scottish Aviation, the idea had first been introduced to the Company by Highland Engineering Limited of Edinburgh, for whom in 1964 SAL produced an initial batch of six vehicles for evaluation, market research and sales promotion. The entire operation from drawing board to delivery was completed in a mere two weeks to achieve a scheduled press and television debut. Highland Engineering subsequently exchanged their association with the project for a royalty, but prospects for the Minatic seemed good after its appearance first at the Kelvin Hall Trades and Industries Exhibition in Glasgow and then at the International Commercial Transport Exhibition in London.

In July 1964 the Board approved production of 360 units at a rate of fifteen per week and simultaneously authorised negotiations to establish a sole distributorship on the basis of a sales programme of 720 per annum.

Again sales were slower to emerge than the Company would have wished, but the 100th unit had been sold by January 1965. One of the intrinsic difficulties of the project was that customers for the Minatic were generally small operators with requirements for small orders, circumstances which demanded disproportionate sales effort. Eventually the project was taken over in 1969 by Cravens Homalloy of Preston, one of Britain's largest builders of commercial vehicle bodies. In the meantime the Scottish Aviation-produced Minatics had lived up to their parentage by carrying loads as diverse as cardboard boxes, dresses, day-old chicks and potato crisps. They were widely used by laundries and furniture removers, but also for applications ranging from mobile office and travelling showroom to the movement of equipment between hospital wards. The Dunlop Racing Service used one specifically to carry a wheel balancing rig to motor races and the GPO used a Minatic as an Outside Broadcast TV Unit.

Of the various non-aircraft ventures conceived by the fertile imaginations of the Project Department in the 1960's the most celebrated was undoubtedly the electric car. The Scamp, a name selected not simply to suggest the combination of Scottish and electric power, but also to reflect the cheekiness of the project, was designed to be a small two-seater urban commuter vehicle. It was felt that for short journeys of this sort it had a number of

The Scamp, a cheeky project

advantages over the conventional internal combustion engined motor car. Off-peak electricity for overnight battery charging was far cheaper than petrol and offered the equivalent of 170 miles per gallon. Requiring only a short range and a limited cruising speed of around 30 miles per hour the electric car could be made very small. Scamp was only 6' 8" long and had a turning cycle of 16', so that it could be easily manoeuvred in tight spaces or even parked end on to the kerb. Other advantages were both environmental and practical. Not only was it quieter than the conventional car but it did not produce toxic exhaust fumes and with only a two-pedal stop-go control system it was not only easy to drive and maintain but did not have the usual difficulties associated with cold starting.

The original design study was followed early in 1965 by the construction of a test vehicle, initially little more than a seat on a chassis, which was quickly christened "the farm cart" around the factory. Despite this unflattering nickname, the test vehicle distinguished itself by rapidly demonstrating that it could do in practice almost exactly what the design study had said that in theory it should do. It was able to achieve a top speed of 36 mph with acceleration from rest to 30 mph in 10½ seconds and the ability to re-start up a one in four incline. In simulated urban driving conditions of six stops and starts per mile it achieved a range of 18 miles against the original target of 15 between re-charging. At a constant throttle setting the range increased to 26 miles.

"The farm cart"

By the end of July 1965 the Company was sufficiently confident with the progress which was being made to approach the Electricity Supply Board, which was known to have been conducting trials with electric cars, with a view to the Scamp being marketed through the Board's 2000 showrooms.

This approach was followed by visits to Prestwick by senior personnel from the Electricity Council and the South West Electricity Board and they were sufficiently impressed for steps to be taken in the Autumn of 1965 to equip the test chassis with a timber, light alloy and perspex body and register it as DAG 908C in order that it might feature in a round of demonstrations in Bristol, headquarters of the SWEB, and in London. This tour culminated in a press conference dedicated to electric cars at which Scamp was driven by Formula One champion racing driver Stirling Moss and was clearly first past the post in generating nearly 200 enquiries from prospective customers and distributors from all over the world.

Encouraged by this reception the Company undertook the manufacture of a batch of twelve pre-production Scamps to be used for further development and market studies. The first of these vehicles left Prestwick at the end of February 1967 to become the central feature on the Electricity Council's stand at the Ideal Homes Exhibition.

Pre-production vehicles

Of the remainder of the Batch, six were allocated to the South West Electricity Board, two to the Eastern and one each to the East Midlands Electricity Board and to the Chloride Electrical Storage Company.

Scottish Aviation had not set out to revolutionise the motor industry. On the contrary the object of the Scamp project had been to use off-the-shelf components and not to require either expertise or research for which the Company did not have the resources. Unfortunately others insisted on evaluating this unconventional vehicle against conventional motor car standards, most notably when the Electricity Council submitted one to test by the Motor Industry Research Association, where it was subjected to the rigours of the pave track. With wheels which were only 8" in diameter it was small wonder that the front suspension broke under this unrealistic punishment. Nevertheless the vehicle was declared unroadworthy and although it was easy enough to strengthen the suspension to cope with MIRA's test track, it was harder to repair the project's punctured reputation.

More fundamental, however, was the problem which has dogged the successful development of an electric commuter vehicle by even the most prestigious organisations right up to the present day, the provision of a satisfactory battery. When Scottish Aviation embarked on the Scamp, electric vehicles such as delivery floats were far from new, but their role was such that they only required low speeds and consequently traction batteries could offer the long life which they simply could not provide to an electric car designed to be an equivalent to the conventional motor car for town driving. Lucas configured a number of sets of special development batteries for Scamp, but in practice the life of these

The alternative commuter vehicle

proved to be limited to around 12 rather than the 18 months which were sought before they could no longer be re-charged but had to be replaced. Lucas were confident that they could develop a battery which would provide the required life but this would have entailed development costs on a scale which was incompatible either with SAL's available resources or with the philosophy behind the Scamp project.

It was also the case that as the Company's aircraft-related business picked up again during the 1960's it turned its attention and resources back in this direction and away from its various vehicle-related activities. This is not to say that it regarded Scamp as a wasted endeavour. Even when the project was finally terminated in 1968 Scottish Aviation did seek to interest other companies in the knowledge which it had acquired, and found a response from firms such as Reliant and Rootes. A couple, at least, of the pre-production Scamps were eventually bought by battery manufacturers in England and Finland and two more ended up working for SAL's Progress Department and for Scottish Express on the loading ramp at Prestwick Airport respectively. The Electricity Board for its part retained one as a museum piece, and another eventually found its way into the historic car collection of Lord Montagu of Beaulieu. Scottish Aviation's own belief that the electric car concept had a continuing future was reflected in the fact that it decided not to sell its patent on the controller which it had developed for the production models, but to seek royalties on its use.

Many other projects were studied but then never undertaken and again a significant proportion sprang from the field of motor transport. The assembly of SAAB cars in this country was for a time during 1959 considered as a possibility for Prestwick. In 1964 a similar arrangement was considered for an Alvis car and the Board was also re-assessing the Project Department's proposals for the manufacture of tipping gear for lorries while at the end of the following year it was investigating the production of a driving simulator. This progressed as far as the assembly of a working prototype before it was dropped when it became clear that several other organisations also had models in production.

Of the various projects which were evaluated but not pursued, the least attractive was almost certainly the manufacture of plastic coffins and the most absurd the production of embossed cigarettes. The most intriguing of the projects which might have been, on the other hand, are inevitably those directly relating to aircraft. Some of these studies dealt with particular peripheral issues ranging from agricultural aviation and reconnaissance pods to an engine-failure procedural training device and an airborne lifeboat which, although not developed by Scottish Aviation, was essentially the basis of the standard UK rescue equipment.

Other projects dealt with complete aircraft and included a surprising number of designs conceived by SAL but never brought to production.

In this respect, as in others, Scottish Aviation was ambitious from an early age. Having only secured its design approvals in 1942, one of the earliest studies which it undertook, even before its proposal to combine Liberator wings and empennage with a new, pressurised fuselage to make the Scottish Aviation "Concord", was for a twin-fuselage, 6-engined landplane and flying boat with a capacity of up to 160 passengers and a maximum take-off weight of 300,000 lbs! From the promotional literature which illustrated Scottish Airlines' plans this aircraft was clearly intended to feature strongly in David McIntyre's plans for the North Atlantic air route and the post-war development of Prestwick.

"The original Concord .... by Scottish Aviation

135 tons to be available either as landplane or flying boat

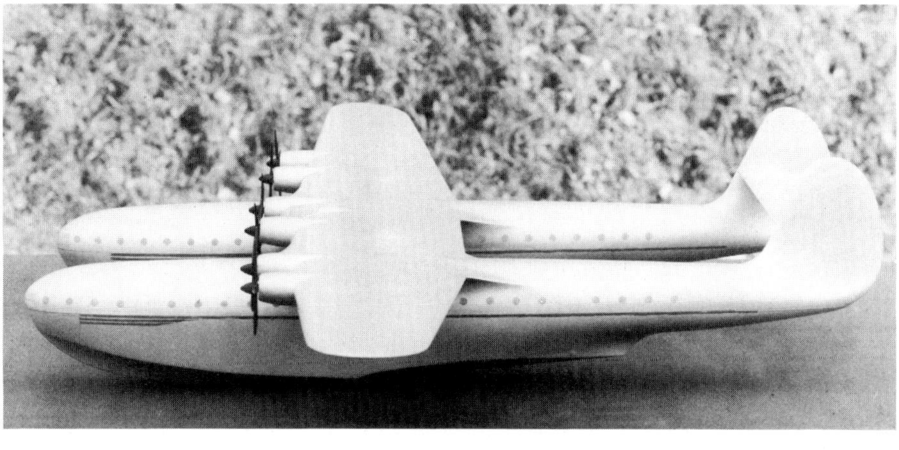

Other early studies included a medium glider of 21,000 lbs maximum take-off weight produced in response to Specification X30/46 and a multi-engine transport aircraft following Specification 26/49, which was aimed to provide a 4-engined replacement for the DH Rapide capable of carrying 10-12 passengers.

The Medium Glider designed to meet Specification X30/46

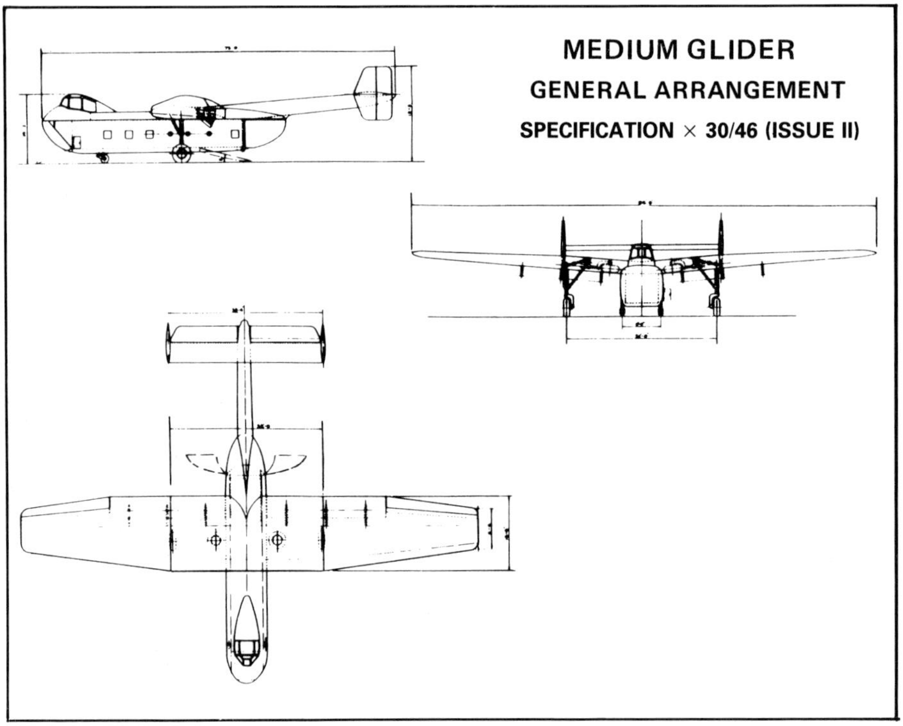

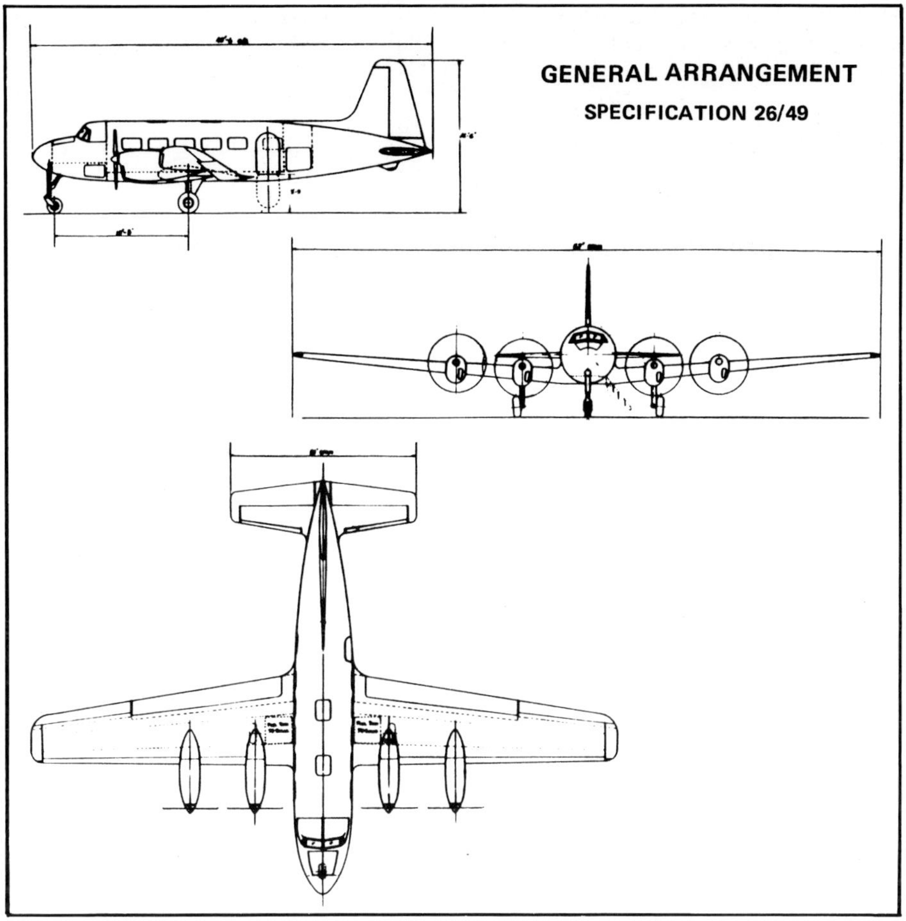

The Multi Engine Transport Aircraft 1949

During the 1950's the design studies were very largely concerned with variants or derivatives of the Prestwick Pioneer and the Twin Pioneer, from relatively modest developments such as the application of floats or of different types of powerplant to more elaborate modifications such as a Twin Pioneer with a rear-loading facility.

The "Turbo-Pioneer", studied between January and April 1956, was intended to be a general purpose aircraft of 31,000 lbs and a capacity of up to 40 passengers. It followed studies made in 1955 for a medium-sized military transport, based on ASROR 323, and was succeeded in its turn by the "Super Twin Pioneer", essentially a "stretched" version of the Twin. This project was pursued from May 1958 to April 1961 with a view to sacrificing some of the Twin Pioneer's STOL performance to achieve better operating economics and in its definitive version it was planned to be a commercial transport of 20,000 lbs for 22 passengers powered by two De Havilland Gnome P1200 engines. The "3-Tonner" studied during 1960 was to be a rear-loading, general purpose light transport of similar weight using the same engines.

The Turbo Pioneer 1956

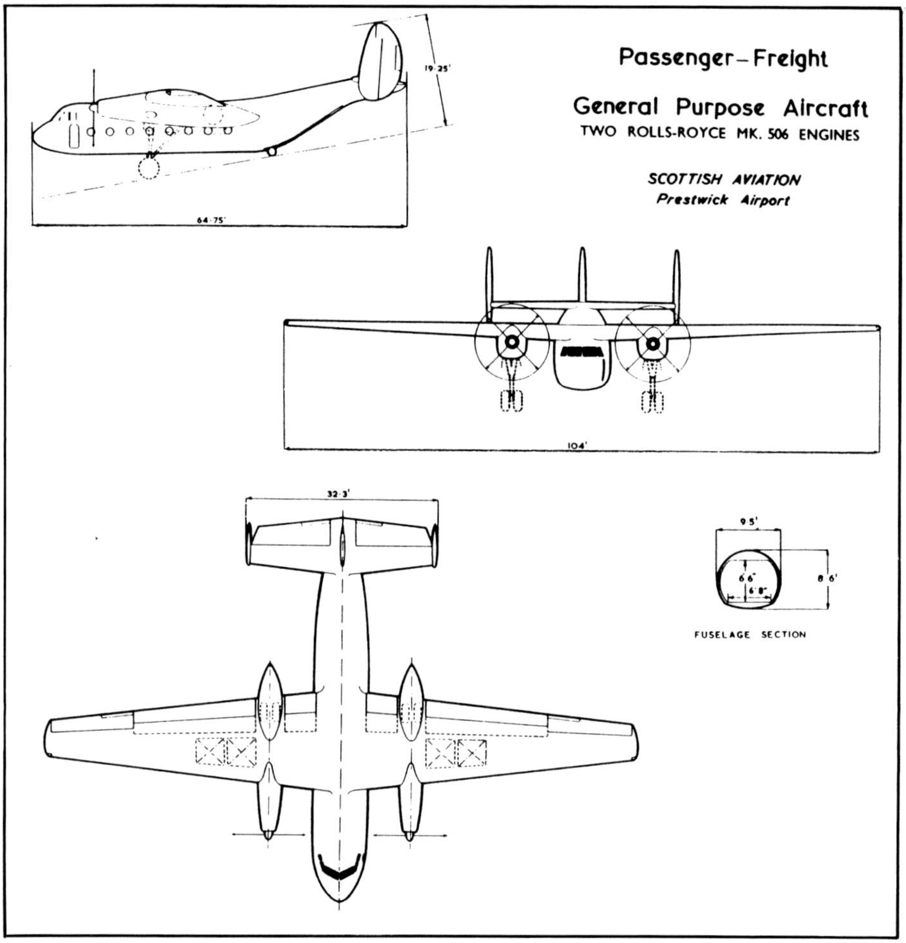

The Super Twin Pioneer Type III c.1960

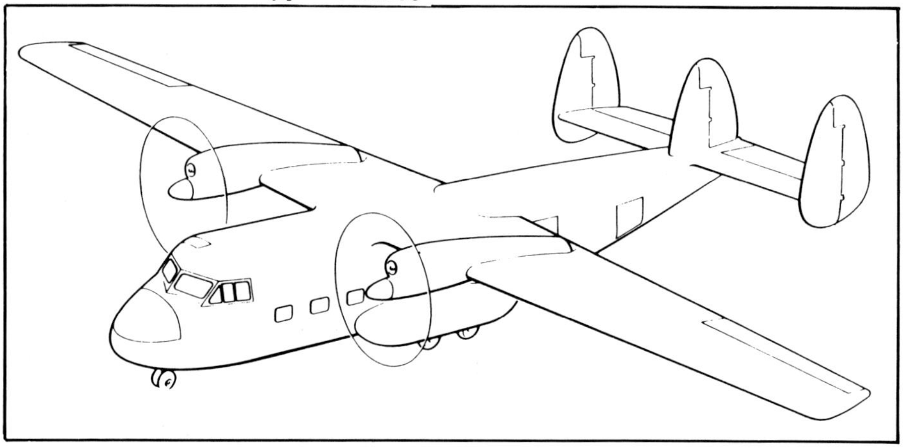

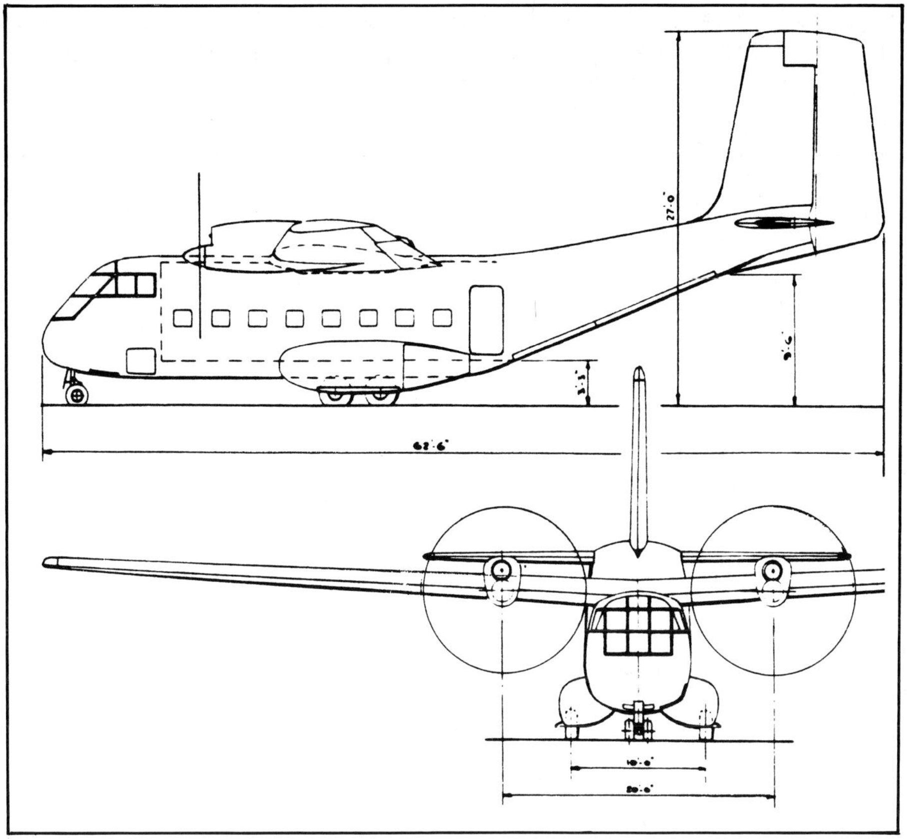

The "3 Tonner" general purpose light transport 1960

The development of the Prestwick Pioneer to equip it with a high capacity fuselage was pursued on and off from early in the 1950's but principally during the period December 1958 to April 1960 under the working title of the General Purpose Light Transport Aircraft. Initially the Type 1 design combined the existing Prestwick Pioneer wings and empennage with a new 8-seat fuselage, but subsequently the scheme of the wings and empennage were reworked in the Type II to arrive at the definitive version of the "Pegasus". A further evolution of the Type I involved the Alvis Leonides Mk 130 (503/7) engine, replacing the 514. This was the aircraft which was to have been named the "Clydesdale".

After the Pioneer projects came a series of studies by the Project Department on three other new designs. The "Gleneagle" was to have been a twin-engined 6-8 seat executive aircraft and was considered in 3 models, using Continental GIO-470-A, Continental GTSIO-520 and Turbomeca Astazou II engines respectively. It was studied early in 1961, as was the "Chieftain" which was to have been a single-engined 6-seat executive and

The Clydesdale, c.1960, evolved from the Pegasus following the Prestwick Pioneer

The Gleneagle executive aircraft, 1961

touring aircraft basically equipped with a Lycoming 1GSO-480-A1A6 but with the same Continental engine alternatives as the Gleneagle.

At the time of their conception these two aircraft had projected selling prices of around £45,000 and £15,000 with, more significantly, associated development costs estimated at £825,000 and £350,000 respectively. It was not surprising therefore that the Company quickly focussed its attention on the third design.

The "Kelpie" was a 2-3 seat single-engined sports and training design which was to have used the Continental 10-360-C engine, and which seemed to address a niche in the private aircraft market which the major American

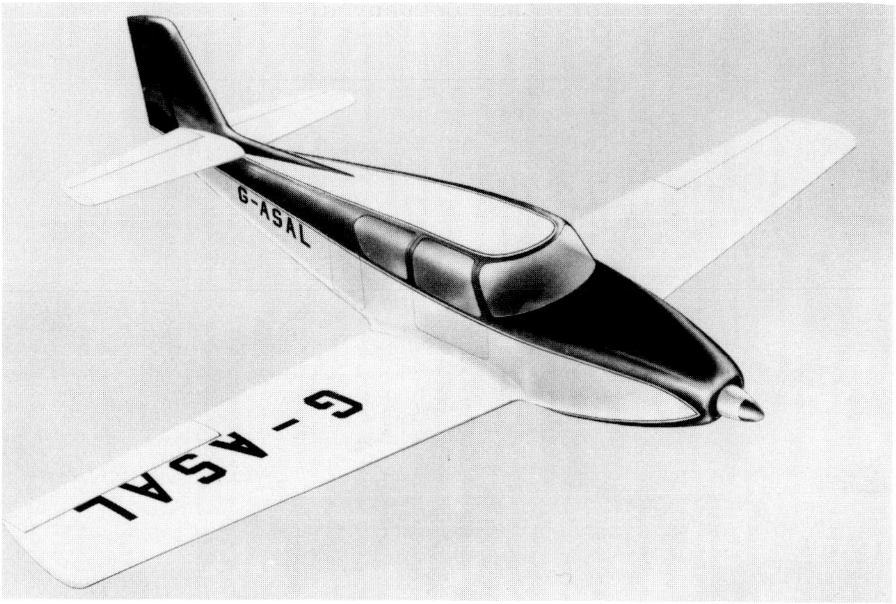

The Chieftain executive and touring aircraft, 1961

The Shelty sports and training aircraft, 1961/1965

manufacturers, such as Piper, Beech and Cessna, were not bothering to tackle. Its proposed selling price in 1961 was £5,000 and the expected cost of its development was a comparatively modest £135,000. Even this was well beyond Scottish Aviation's resources in the context of its financial reconstruction after the Twin Pioneer and the

Kelpie too was therefore dropped, although The Mooney Corporation in the United States expressed interest in manufacturing it and studies went on throughout the period 1961 to 1966 during which time the name of the aircraft was changed after it had been pointed out that the Kelpie in Scots folklore is a water-sprite with a reputation for a malicious disposition. The aircraft was duly re-christened the "Shelty" after that much more attractive creature the Shetland Pony.

Subsequent design studies on complete aircraft initiated by Scottish Aviation were principally variants on the Bulldog and Jetstream after the Company had taken over the design rights of these aircraft from Beagle and Handley Page respectively. Two other names of aircraft which might have been however also deserve a mention because they illustrate the fact that, even after it had acquired the Bulldog and the Jetstream, Scottish Aviation did not abandon the pursuit of its own designs, the commonly accepted summit of the 'plane-maker's art.

The "Don" was the name given to the Type I of a study carried out between May 1974 and November 1975 into a basic pilot trainer. The study ran to eight distinct types in all but the "Don" was designed to be a 2/4 seat multi-role light turbo-jet and as such it was a direct antecedent of the last of the original Scottish Aviation design projects.

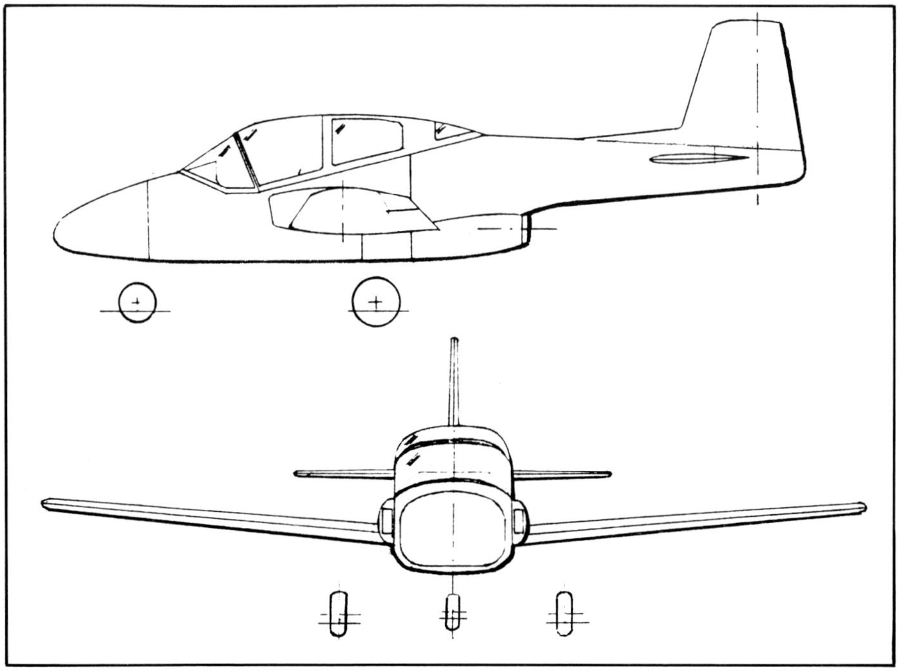

The Don Type I multi-role turbo jet, 1974/1975

The "Sapphire" was studied between 1975 and 1980 as the definitive Scottish Aviation concept of the basic Pilot Trainer powered by a Lycoming ALF101G. It was aimed at fulfilling various roles, including strike attack as well as training , and was consequently known by the inelegant title of "LUMPS" or Light Utility Multi-Purpose System. Other labels applied to particular functions of the same project, in an industry with a positive mania for cryptic abbreviations, included "LCTS" and "BUSS", or Low Cost Training System and Battlefield Utility Support System respectively.

The Sapphire 1100, light utility multi-purpose jet, 1975/1980

"Sapphire" was directed towards an identified world-wide market need for a new basic trainer and in 1977 it became the subject of a detailed specification which showed that with its high by-pass fan engine it would be capable of 300 knots at sea level and a very rapid rate of climb. Following the Company's absorption into British Aerospace, Jetstream 31 was given preference over "Sapphire" as the new project for Prestwick, but the design studies which Scottish Aviation had carried out were subsequently added to the work being done at Brough and in this way SAL made its own direct contribution to the design activities of the new corporation.

If studies for the Company's own designs were numerous, then evaluations of participative projects through which Scottish Aviation could be involved in other organisations' aircraft were legion. The list of participation projects studied after the formation of the Project Department in 1963 is formidable and is worth citing at length:

Fraser-Walker Gyro Jet
Aircoupe
Fairtravel Linnet
Miles Century/Student/Graduate
Malmo Vipan
Riley Dove "400"
Victa Airtourer/Aircruiser
Handley Page Jetstream Mk 1 and Series 200
General Aircraft Corporation 100
Komov KA-26
Camco V-Liner
Hoverhawk Hovercraft
Britten Norman Islander
American Aviation Yankee
Beagle Pup, 206 and Bulldog
Wing Aircraft Derringer

Land Development Aircraft (Lockspeiser)
Beech Aircraft
Cierva Twin Helicopter
Wassmer Aircraft
Avions Pierre Robin
IAR-823
Shrike Commander
Light Observation Aircraft
Bede BD-5
AM-C111
Lündberg Project
AR404
Turbine-powered Ag-Wagon
RFB Fantrainer
Saab Transporter
Practavia Sprite

Among the most interesting features of many of these potential projects was the fact that commonly it was the external organisation which approached Scottish Aviation seeking a collaboration rather than vice versa. The

Board Meeting in November 1961 for instance considered two approaches. The first came from Procaer, an Italian manufacturer of sophisticated single-engined touring aircraft, which was looking for SAL to become its UK agent. The second was from Aero-Commander Inc of Oklahoma with a view to Scottish involvement in assembly or sales of its 500A aircraft.

Although the Company chose not to pursue either of these opportunities, it was not long before other approaches were forthcoming. In 1964 Miles Engineering pursued SAL to take an interest in its proposed Executive Twin-Jet, the "Century". In 1966 the Company was invited to take up the manufacture of the Victa "Airtourer" to be followed by the "Aircruiser" and was simultaneously being sounded out with a view to the production of components for the Britten-Norman "Islander". The most remarkable association which was put forward came in 1968 when Soviet officials explored the possibility of Scottish Aviation taking on the manufacture and sales of the Russian KA-26 Helicopter.

It is clear then that despite the uncommon variety of projects, which the Company explored in its pursuit of diversification and as a matter of expediency in recession, its inclination was always to return to the business for which it had been established. It is also evident that despite its varying fortunes, Scottish Aviation had retained a valuable reputation in the industry. In these circumstances it was perhaps inevitable that it would only be a question of time before the Company was again engaged in the manufacture of its own aircraft.

Bristol University Air Squadron flying Scottish Aviation Bulldogs.

Chapter 9

"SALvage" Changing Fortunes

In the financial reconstruction of the Company which followed the Twin Pioneer project the Finance Corporation for Industry found itself, strictly contrary to its terms of reference, in the position of being the major shareholder in Scottish Aviation. By January 1965 it was actively looking for a buyer to relieve it of this obligation and, although the earlier notion of amalgamation with another aircraft manufacturer was not pursued, a number of companies outside the industry itself expressed interest in acquisition.

The most unexpected was British Eagle Limited, which was at that time the second largest independent airline in Britain. In pursuit of a policy of expansion it was seeking a maintenance base outside London for its fleet, a role for which the facilities and skills which SAL had to offer at Prestwick Airport seemed ideally suited.

Other prospective buyers at this time were Solar Industries, Alvis and Lindustries Limited, all essentially interested in Scottish Aviation as a possible means of adding to their existing engineering capabilities.

Lindustries in particular were still actively interested when a new name came into the reckoning. Cammell Laird and Company based in Birkenhead was a long-established shipbuilding and engineering group which had been building ships for over 135 years. Among the distinguished vessels which had been launched from its slipways on the Mersey were the first ever all-welded ship, the merchant vessel Fullagar, and a series of famous Royal Naval ships including two Ark Royals and the Prince of Wales.

By November 1965, when this unexpected foster parent first expressed interest in taking over Scottish Aviation, it was already clear that the British Shipbuilding industry was coming under increasing pressure from foreign competition and Cammell Laird was consciously pursuing a policy of diversification away from the declining sector of British naval work into merchant shipbuilding and simultaneously broadening its range of business through a series of acquisitions.

Events, which had moved slowly since the Finance Corporation for Industry first declared its intention to find a buyer for SAL, now gained momentum and the purchase was complete by the end of April 1966 when it was announced to a surprised world. The secrecy which had surrounded the transition had, however, had the very beneficial effect of averting the uncertainty and discomposure which so frequently accompany take-over situations. In this case the outcome of the change in ownership was wholly positive because it meant that Scottish Aviation was now part of a group which was actively looking to make money available for investment in new projects.

The smoothness of the transition owed much also to the fact that there was no managerial upheaval in the acquired company, another typical product of the acquisition process. Cammell Laird appointed a new Chairman, Sir Leonard Owen, known as "Owen the Bomb" as a result of the sixteen years which he had spent as a leading executive on the Atomic Energy Authority, and SAL was duly briefed on Cammell Laird's particular reporting and accountability procedures, but otherwise the Prestwick management was given its head. Scottish Aviation had been taken over very much as a going concern and its acquisition was an expression of the new owner's confidence in it.

The action which perhaps best illustrated that this was the basis of Scottish Aviation's incorporation into the Cammell Laird organisation was the parent company's prompt decision to purchase a De Havilland Dove to be based at Prestwick in order to provide an executive air service to the whole Group.

New Aircraft, new Chairman : the Dove and Sir Leonard Owen.

The Cammell Laird take-over gave SAL the opportunity to demonstrate its potential and, most immediately important, it provided the source of finance to enable the Company to realise its objective of entering into a new aircraft project as a major sub-contractor. The venture in question was the launch by Handley Page Limited of a design which aimed to provide a highly competitive new generation small airliner. This was the HP.137, otherwise known as the Jetstream.

The possibility of Scottish Aviation being involved in this programme was first explored in September 1966 when the Chairman and Managing Director of HP visited Prestwick. The visit and the subsequent discussions were evidently satisfactory because by November a provisional order had been placed on SAL to manufacture 54 sets of wings for the aircraft with delivery of the first two sets to take place in October 1967.

Wings for the Handley Page Jetstream in production at Prestwick, 1969.

Handley Page had in fact been so impressed with the Scottish Aviation facilities that it pressed Cammell Laird for some time in an attempt to acquire a shareholding in the Company so that it might ensure that more of SAL's capacity could be diverted from its existing range of projects and applied to work for Handley Page, which was projecting that demand for Jetstream would soon necessitate production at the staggering rate of 15 aircraft per month. The prospect of a dependence of this sort was, however, acceptable neither to the Cammell Laird nor the Scottish Aviation Boards and it was agreed that the Company's relationship with HP should be confined to that of sub-contractor.

It was a prudent decision because, despite its long and distinguished history of aircraft production and despite the extremely buoyant market which it forecast for its new commuter airliner, particularly in the United States, Handley Page itself was faced with considerable difficulties by 1966. Originally incorporated in 1909, HP was the first Limited Company to have been formed in Great Britain to manufacture aircraft. Sir Frederick Handley Page and, after his death in 1962, his successors had proudly guarded the firm's independence, refusing to allow it to be drawn into the process of amalgamation which had progressively absorbed so many other famous names into the two major groups, Hawker Siddeley Aviation and the British Aircraft Corporation. The Government, committed to this rationalisation of the industry, responded by making it clear that Handley Page could not expect to receive new design work nor, by implication, the financial assistance necessary for the large aircraft on which the company had traditionally worked.

The HP.137 was therefore designed to fit a carefully identified gap in the feeder liner market, bringing large aircraft sophistication to the production of a small aircraft which would be compatible with the company's funding capability. The result was a pressurised 18-seater powered by two turboprops, for which, in the absence of a suitable British or American engine in the required power range, the French Turbomeca Astazou XIV was selected. These 850 hp engines with their reversible pitch propellers were expected to give the 12,500 lb aircraft a cruising speed of 300 mph over a range of 2,000 miles.

The Jetstream was aimed principally at the United States market where a dealer network was established even before the prototype, G-ATXH, made its first flight from Radlett on 18th April 1967. By 1969 the distributor arrangements suggested that advance orders could be as high as 181 aircraft and HP had therefore tooled up to produce at least 500 units. Paradoxically the strength of this interest itself contributed to Handley Page's undoing. An ambitious development and production programme was set and almost immediately began to slip, while simultaneously the pace and scale of the programme imposed massive financial demands.

In consequence Scottish Aviation found itself subject to parallel pressures. Undertaking the manufacture of the HP.137's mainplane had required two processes to be introduced to Prestwick for the first time. The chemical milling plant and the bonding autoclave were both developments which enhanced the range of manufacturing capabilities which the Company had to offer for the future, but in the meantime they were major items in a total expenditure which was growing alarmingly quickly while the prospects of returns were coming no nearer.

Arrival of the autoclave for the wing skin bonding process.

The Scottish Aviation Board grew increasingly concerned during the summer of 1969, but allowed the Handley Page debt to continue both because it believed that the Jetstream aircraft as such was a viable proposition and also in anticipation of receiving a large share of work from HP on the impending conversion of its Victor bombers to an in-flight refuelling tanker role. On 7th August, however, other creditors forced a bankrupt Handley Page into receivership.

At this point Scottish Aviation's investment in the Jetstream contract amounted to £2,200,000 and the Receiver, Mr Kenneth Cork (subsequently Sir Kenneth Cork, Lord Mayor of London) suggested that SAL stood to gain more by Handley Page continuing to operate than it would from an immediate liquidation. In this way Scottish ironically found itself in precisely the same situation as its own sub-contractors had done with the Twin Pioneer project a decade before.

In fact the Company had already taken steps to organise itself to manage this situation. In May T.D.M. Robertson was appointed Deputy Chairman and devoted himself full-time to the job of protecting and recovering Scottish Aviation's investment in Handley Page, becoming Acting Managing Director at Radlett, while Hugh Laughland, formerly Executive Director, Finance, took over the responsibilities of Managing Director at Prestwick, throwing his considerable personal energy into this new task.

SAL's preference, naturally, was to find some way of ensuring that the Jetstream project could continue. Although the Astazou engines were failing to achieve the specified economic performance, nobody seemed to have any doubt that the aircraft had potential. Even in receivership Handley Page's reduced rate of production called for five sets per month. There was considerable optimism therefore when an American named K. R. Cravens indicated at a meeting in London that his corporation would be prepared to inject £3,000,000 into the project. This lifeline, however, was cut early in 1970 when Cravens himself became the victim of a terminal illness.

The Jetstream production line closed at Prestwick on 27th February 1970 and the prospect of Victor work now seemed remote. Handley Page had come to terms with the choice of "amalgamation or death" far too late. Neither of the major aircraft groups had any further interest in taking over HP since it was clear that any work which the Government might have directed to Handley Page would now come to them in any case. In April 1970 Handley Page went into liquidation.

Unfinished Jetstreams at Radlett after the collapse of Handly Page in 1970.

From the ruins Scottish Aviation managed to acquire the spares support responsibility for the Handley Page Dart Herald and also the factory at Cumnock which had previously been Handley Page (Scotland) Limited. At the same time SAL was left with a redundancy situation and with twenty-two sets of Jetstream wings in various stages of completion and apparently no prospect of being able to realise this investment.

Unfortunately Cammell Laird itself was in no position to provide the funding to enable Scottish Aviation to take on the Jetstream as a total project because the Handley Page collapse had coincided with difficulties of its own in the shipbuilding business. The consequences were critical.

The problems sprang from the advanced merchant vessel work which Cammell Laird had pursued after 1968 when the Government had announced that no new orders for nuclear submarines could be expected. The new merchant ships required considerable development with the result that when delays ensued and both material and labour costs rose faster than had been expected the vulnerability of the company's fixed price orders and modest escalation clauses were swiftly exposed. A specially prepared auditors' report in March 1970 revealed that it was already inevitable that the shipyard would make a loss of approximately £10 million on its £44 million order book. It became clear that the Birkenhead yard needed substantial modernisation and capital investment to equip it for its new type of work, but irrespective of this it was estimated that, as a result of the shipyard's losses, the CL group as a whole would require to find injections of cash amounting to £4 million in 1970, some £5 million in 1971 and another £1.5 million in 1972. Without immediate action the group faced the prospect of being unable to meet its wage bill from May 1970.

Cammell Laird looked for assistance to the Shipbuilding Industry Board, a statutory organisation which had been established in 1967 to stimulate the reorganisation of the industry and to administer the public funds made available for that purpose. Although the SIB had the power to make loans to, or even to buy shares in, shipbuilding firms, it assessed the Cammell Laird situation as one with which it would prefer not to become involved.

CL therefore appealed to the Government and the Paymaster General, Harold Lever, invited another statutory body, the Industrial Reorganisation Committee, to come to Cammell Laird's rescue. Cammell Laird and Handley Page had in fact jointly approached the IRC in 1969 in an unsuccessful attempt to obtain its support for HP. The IRC's investigation of CL now left it extremely concerned that the prospective losses on the shipyard's outstanding orders were virtually unlimited. It also felt that Cammell Laird's shipbuilding activities were properly the concern of the SIB and the IRC was not prepared to put its money at risk by stepping in directly. The solution which was eventually devised therefore saw the Government take up a 50% equity share in the shipbuilding company which was divorced from the remainder of the CL group so that it alone would be responsible for any further losses which it incurred.

The other companies from the old CL were reorganised in 1970 into the Laird Group with a young IRC executive, John Gardiner, as the new Managing Director and Chief Executive. Separated from the shipyard, the new group's financial performance went from strength to strength. By 1977 group turnover had doubled and pre-tax profits had increased by 450%.

With a new Laird in both senses of the word Scottish Aviation was therefore in a fortunate position at a time when other more celebrated firms were to be counted among the victims of the spate of corporate collapses which took place early in the 1970's. In the aircraft industry the disappearance of Handley Page was followed by the dramatic failure of Rolls Royce, in which SAL again found itself in the position of anxious creditor. At much the same time, however, it became apparent that the clouds of collapse could, for Scottish Aviation at least, have a silver lining.

The silver lining took the form of an aircraft which was to become almost synonymous with Scottish Aviation during the 1970's. This aircraft was the Beagle B.125 or Bulldog, a military development of the successful Pup.

Original Beagle Bulldog prototype with experimental 3-bladed propeller.

If Scottish Aviation had taken up the earlier opportunities which had presented themselves during the 1960's to become associated with Beagle, then it might in 1970 have found itself again in the position of a stranded sub-contractor. As early as June 1961 exploratory discussions had been taking place between the two companies with a view to SAL undertaking manufacture of the wings for the new Beagle 206. But the design required that the wings should be bonded rather than rivetted and SAL estimated that it would need a production run of 200-300 aircraft to recover the cost of investing in

Redux-bonding plant. In these circumstances it chose not to pursue the prospective order for 150 sets.

The story did not end there because in subsequent discussions late in 1962 Beagle enquired whether SAL would be interested in building the complete aircraft other than the wings and ailerons which had by this time been sub-contracted to Boulton-Paul Aircraft Limited. Negotiations continued through most of 1963 but Scottish reckoned that without the backing of a substantial military order the aircraft was unlikely to sell in significant numbers and therefore declined to participate on a risk-sharing basis. In October Beagle decided to proceed with the project in its own factories and consequently there was no sub-contract work for Prestwick.

For a time it appeared that SAL might have cause to regret its decision to stand off. In 1964 the Company was given to understand that all tenders for small aircraft which were not of interest to HSA or BAC would be placed with Beagle. Scottish Aviation was quick to remind the Government of its own attributes in this particular field.

Beagle itself seemed to have decided, however, that the range and nature of Scottish Aviation's facilities provided a useful complement to its own and in March 1966 its Managing Director, Peter Masefield, again approached SAL with a view to the two companies becoming production partners. By this stage Scottish Aviation's own capacity was fully committed and these circumstances, coupled with its continuing concern about Beagle's viability, made it easier to arrive at the decision not to take part.

The doubts about the future of the Beagle Aircraft Company proved to be well founded. The original Bulldog prototype, G-AXEH, converted from a Pup, made its first flight on 19th May 1969 but Beagle never completed either the development or a production aircraft. Early in 1970 the firm went into receivership.

Yet it left an important legacy. Before its collapse Beagle had secured a major contract for 58 aircraft from the Royal Swedish Air Force, an order which was the more notable for having been achieved in the face of tough competition from Sweden's own MFI.17. Given this insight into the quality of the Bulldog as an aircraft, Scottish Aviation was quick to enter negotiation with the receivers, as a result of which, in May 1970, with the agreement of the British and Swedish Governments, SAL became responsible for fulfilling the RSAF order and acquired the design, production and marketing rights for the Bulldog plus product support and design responsibility for the Pup, the Basset and the Beagle 206.

Bulldog in Royal Swedish Air Force livery.

The Bulldog was essentially a 2/3 seat primary trainer with an Avco Lycoming 10.360 200 horsepower fuel-injected engine. Among the particular features which had attracted the Swedes were the spaciousness of the cockpit, its wide field of vision, the aircraft's combination of safe flying and handling characteristics - notably straightforward stall recovery together with a full aerobatic capability - its suitability for instrument training and its durable fatigue life. During the first few months of 1970, even before the project was taken over by Scottish Aviation, the aircraft underwent extensive type evaluation in Sweden. These trials included the use of a ski undercarriage and of wing-tip pylon-mounted missiles to assess the aircraft's suitability in a ground support role.

Swedish Bulldog on skis.

G–ASAL, Bulldog demonstrator equipped for rocket trials.

These operations had two important results. The first was an extension of the Swedes' order to convert twenty of their options into firm orders for the Swedish army. Secondly, and more generally, they meant that relatively little design, aerodynamic or engineering development activity remained to be carried out on the aircraft when it was taken up by Scottish Aviation. Consequently the Company was able to apply itself singlemindedly to the task of establishing the jigs and tools necessary for production, since although some 150 Pups had been built and sold, the Bulldog project had never reached this stage under Beagle.

The prospect of once more producing an entire aircraft of their own propelled the efforts of everybody associated with the project at Prestwick. The first production aircraft G-AYWN made its maiden flight on 22nd June 1971 and the first delivery in the demanding timetable set by the Swedish contract was achieved on schedule on 26th July that same year.

At the time of acquiring the rights to the Bulldog the Company believed that it could reasonably expect to sell in excess of one hundred aircraft and this confidence was quickly borne out by the repeat orders from Sweden and by orders from Kenya and from a satisfied customer of Scottish Aviation's in the past, the Royal Malaysian Air Force.

However, the prize which the Company was most anxious to win, not only for the enhanced reputation which it could be expected to bestow on the aeroplane in the eyes of overseas customers, but also because it represented a requirement for 132 aircraft in its own right, was the contract to supply the Royal Air Force with a new basic trainer to replace the Chipmunk. Before the end of 1970 SAL was duly putting forward the Bulldog's claims to meet this requirement and after the contract was secured production rose to a rate of 6 or 7 per month and sales went on to total 325 aircraft spread "the world o'er". Operating in Sweden, Malaysia, Kenya, the United Kingdom, Ghana, Nigeria, Jordan, Lebanon and Hong Kong, the Bulldog easily outsold all its competitors.

First production Bulldog taking off on its maiden flight.

The Bulldog was not the only salvage operation which SAL undertook in 1970. Late in the same year an RAF requirement for a multi-engined pilot training aircraft was identified and so began 18 months of negotiation as Scottish Aviation pursued this opportunity to revive the Jetstream, having acquired the rights to the aircraft from Jetstream Aircraft Limited, the company which had been formed by American entrepreneur Bill Bright in an attempt to pursue the project in the aftermath of Handley Page's demise.

A military version of the Jetstream had in fact been projected by Handley Page after it had obtained a contract for eleven aircraft for the United States Air Force against stiff competition from American manufacturers. Designated Mk.3M, or by the USAF the C-10A, the prototype G-AWBL made its first flight on 21st November 1968. The most significant feature of this variant of the aircraft was its new powerplant, the Garrett AiResearch TPE-331-3A-301W turbo prop, which made it the precursor of the British Aerospace Jetstream 31.

Although the United States Air Force order had lapsed with the demise of Handley Page, the fact that it had been awarded in the first place was powerful evidence of the aircraft's suitability for a military role. During the week between Christmas and Hogmanay in 1971 Scottish Aviation's thoughts were concentrated not on the forthcoming New Year's celebrations but on the final

R.A.F. T Mk.1 Jetstream trainer.

preparation of its tender for the RAF multi-engined pilot trainer contract which had to be in the hands of the Ministry of Defence by 29th December. Certainly the real New Year and cause for celebration at Prestwick in 1972 came on 24th February when the Government's decision to award the contract to Scottish Aviation's Jetstream was announced in the House of Commons.

The Jetstream T.Mk.I programme for the RAF called at this stage for 26 aircraft to be completed by the end of July 1974. From the Company's point of view the contract was, in the first instance, an opportunity to recover the funds which it had sunk in the Jetstream while it had been a sub-contractor to Handley Page. Already in stock either at Prestwick or Sywell were fifteen complete sets of wings, while "work-in-progress" at Prestwick accounted for a further seven sets at various stages of production. Three fuselages were complete but requiring overhaul and conversion to RAF requirements, as did a further eleven which had been broken down for storage. Seven were partially completed and the remaining five therefore involved new manufacture.

It was established at an early date that North West Industries of Canada, which had been responsible to Handley Page for the supply of the empennage, would fulfil this same role for the T.Mk.I, but the question of who would build the fuselages, previously produced by HP itself, was more problematic. During 1971 discussions took place with a view to the work being performed by the British Aircraft Corporation, and Short Brothers of Belfast also offered to do the job, but the quotations received were judged by Scottish to be too high. SAL therefore determined to build the fuselages itself and committed itself to the revival of the Jetstream not merely as a means of recouping its earlier investment but also as an opportunity to prove the aircraft's potential in the civil market which Handley Page had originally perceived as being its principal application.

Finding a new home at Prestwick.

1973 therefore was in many ways a vintage year for Scottish Aviation. On 12th February it was appropriate that it should be the Company's President, the Duke of Hamilton, who was responsible for the handover to Air Vice Marshal Cook, Director General Training, of the first of the 132 Bulldogs for the Royal Air Force, most of which were destined for the Central Flying School at Little Rissington and the sixteen University Air Squadrons. It was an occasion which recalled Scottish Aviation's long association with RAF pilot training since its very earliest days as the Scottish College of Aviation in 1935.

The Duke of Hamilton hands over the first RAF Bulldog to Air Vice Marshall Cook, 12.7.73

Chairman Sir Maurice Banks hands over the first RAF Jetstream to Air Chief Marshall Wheeler, 26.6.73. H.W. Laughland is on far left and T.D.M. Robertson on far right.

Another important milestone was attained in April when on Friday 13th superstition was laid aside and the first Scottish Aviation production Jetstream took to the air for its maiden flight. This was quickly followed on 26th June by the second major ceremony of the year when Sir Maurice Banks, Chairman of the Laird Group, handed the first of the Royal Air Force Jetstream aircraft over to Air Chief Marshal Sir Neil Wheeler, Controller of Aircraft for the Ministry of Defence.

It was a year in which Scottish Aviation had good cause to congratulate itself. Not only had it weathered the economic storm at the turn of the decade which had seen so many famous names disappear but it had successfully re-established itself as a prime contractor in aircraft manufacturing and was well on the way to making the name Scottish Aviation once again synonymous with the idea of RAF training.

Employing over 2500 people it had a balanced spread of programmes which included major sub-contracts for customers both at home and abroad. The Royal Canadian Air Force overhaul and maintenance contract was entering its 20th successive year, work for the Lockheed Corporation in the United States had been extended from C.130 fuselage panels and pylons to include work on the Tri-Star and overhaul was ongoing for Rolls Royce, Pratt and Whitney and Avco Lycoming engines.

Lockheed Tristar doors in production

Furthermore it appeared that this foundation of long-established associations could be expected to expand through the addition of new programmes, such as the work which was undertaken for the Hawker Siddeley Group on the Harrier VTOL fighter and the Trident airliner. There was also a prospect of involvement in the new HS.146 project. Discussions in this connection during 1973 resulted in the contract for the manufacture of the 146 air brake being awarded to Scottish Aviation in February 1974.

But 1973's successes were also tinged with sadness. On March 30th, less than two months after he had presided over the handover of the first RAF Bulldog, Group Captain the Duke of Hamilton and Brandon, co-founder of Scottish Aviation, died suddenly in Edinburgh at the age of 70. Although he had relinquished the Chairmanship of the Company in 1961, he and his family had always remained closely and actively interested in SAL's affairs. His eldest son, Angus, Marquis of Clydesdale, had worked as one of the pilots on the Bulldog programme after leaving the RAF. The Duke himself, attending the Board Meeting on 26th April 1972, described in detail to his colleagues the Royal Netherlands' Air Force's future requirement for a light trainer and the steps which he had initiated to introduce the Bulldog as a possible contender through arranging an opportunity for His Royal Highness Prince Bernhard to fly the aircraft in person. He was a man who was not content to be an observer. As a young man he had working incognito as "Mr Hamilton" at the coal face of one of his family's mines and had held his trade union card. Throughout his life Douglas-Hamilton had a passionate interest in Scotland, in aviation and in the creation of employment, interests which were combined and manifested in Scottish Aviation Limited, interests which were rewarded by the Company's achievements. Appropriately it was from the cockpit of a Scottish Aviation Bulldog that his ashes were scattered to rest in his native land.

The Bulldog production line, 1971

"It takes one to know one"

The death of its founder was a reminder to the Company that it lived in a state of ever-present change. If a further reminder were needed of the changing fortunes of any organisation in the aircraft industry, then it was not long in coming. Within a few months of 1973's triumphs the Company found itself confronted by three major pressures.

The first were economic difficulties precipitated by the raging inflation which followed the 1973 oil crisis. Apart from the obvious general effect which this had on the Company's trading it contributed directly to the Government's decision to review defence expenditure and meant, so far as Scottish Aviation was concerned, that the prospect, for which the Company had hoped, of additional Jetstream orders from the Ministry of Defence simply disappeared.

The second pressure also sprang the problematic economic environment and took the form of a general recession in the aerospace industry world-wide. Inevitably this hit hardest those firms, such as Scottish Aviation, which depended heavily on sub-contract orders from the major manufacturers. Hawker Siddeley, for instance, faced with the prospect of rapidly escalating costs on the HS.146 project, decided to cancel the programme and Scottish Aviation's work on the air brake was therefore lost.

The third pressure was perhaps the most difficult because it was new and undefined. This was the uncertainty which was caused by the prospect of the aircraft industry being nationalised. The plan to bring the aircraft industry into public ownership and control had been set out in detail in a joint statement by the Labour Party, the Trades Union Congress and the Confederation of Shipbuilding and Engineering Unions and was reaffirmed as policy in the Labour Party Manifesto before the 1974 General Election.

When Labour won the election, the effect on the industry was immediate. Overseas customers of Scottish Aviation were quick to write seeking indications of how the legislation which now seemed inevitable would affect the Company's ability to perform. "Will it have," asked one letter directly, "any deteriorating effect in the areas of schedules or costs?" Clearly the reputation of Britain's nationalised industries at that time was poor and the prospect of nationalisation certainly seemed to Scottish to have an inhibiting effect on foreign customers' willingness to enter into new commitments, at least until the implications of public ownership of the industry could be assessed in practice.

Of course the existing parent companies in the industry had their own views regarding nationalisation. The Society of British Aircraft Constructors opposed the proposal strongly. One of the features of the original Labour Party plan, which was most worrying to the industry, was the suggestion that the compensation made to former owners had, in the case of previous nationalisations, been more generous than it was now intended to be with the aircraft industry on account of its heavy reliance on inputs from public funds and on the Government as its principal customer. The SBAC countered this proposal by arguing that public funds injected into the industry had been more than offset by the benefits which had accrued to the nation in terms of technical spin-offs, defence, the balance of trade, employment and taxation.

In any event, although the Aircraft and Shipbuilding Industries Bill stipulated that any new money invested in the Company after February 1974 would either be repaid in full or compensated with the same effect, there is no doubt that in the case of Scottish Aviation at least the possibility of nationalisation served to limit the new investment which the parent company was prepared to make.

The original nationalisation proposal aimed to bring together the giants of the British industry, the British Aircraft Corporation, Hawker Siddeley Aviation and Hawker Siddeley Dynamics. Scottish Aviation was not included in the initial scheme. Scottish Aviation's Management had mixed views on this situation. It did not relish the

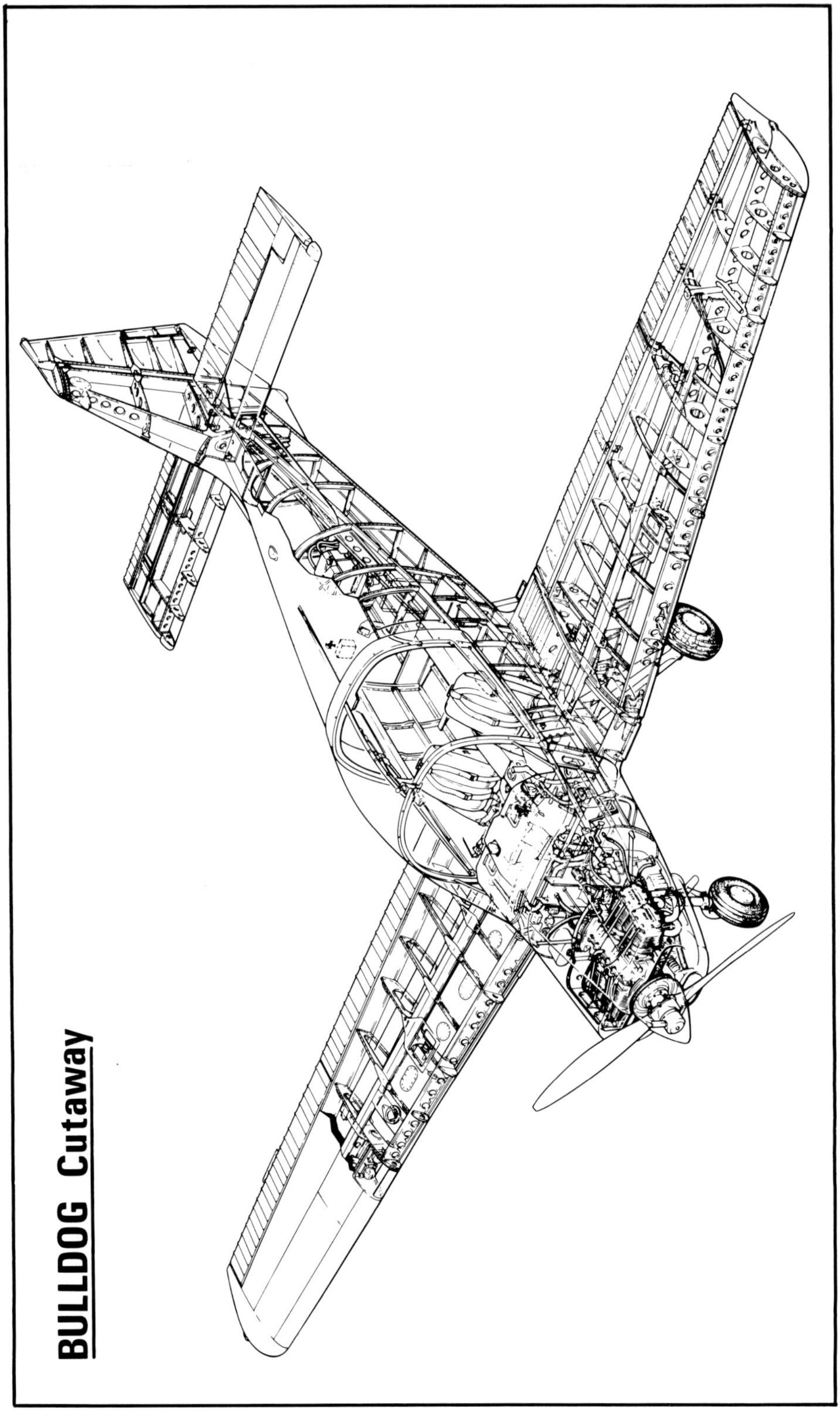

The Bulldog general assembly.

prospect of losing its independence to govern its own affairs but at the same time it had to ask itself whether the Company could expect to survive outside the nationalised body.

It seemed, for instance, that the new public corporation could be expected to have the benefit of interest-free capital and might not be bound by the same requirement to make profit as Scottish Aviation, which, remaining in the private sector, would continue to be responsible for generating the positive cash flow necessary to secure its future and development.

" Taken under the wing ", Bulldog sizing up to a Lockheed Galaxy

The Company was also concerned that the nationalised corporation might decide to draw back work which its predecessor companies had previously sub-contracted to Scottish and equally that work for foreign countries arising from offset arrangements, like the Lockheed C.130 work which had derived from the RAF Hercules purchase, might subsequently cease to be open to competitive tender but simply be directed by the Government to the nationalised manufacturing company.

Making its representations to the Department of Industry in January 1975 Management of Scottish Aviation and the Laird Group pressed both for a specific instruction to be placed on the public corporation requiring it to allow SAL to compete for its work and also for specific limits on how small an aircraft the nationalised company would be allowed to produce, the aim being to protect the particular segment of the aircraft market in which Scottish had specialised.

Trade Union representatives and Members of Parliament were swift to express their own concern on behalf of members and constituents that the dominant position of the proposed nationalised corporation would sign the death warrant of Scottish Aviation as an employer. For their part they were clear that the only solution was for Scottish Aviation to be incorporated into the new public body. It was very largely as a result of the pressure which they were able to exert towards this end, coupled no doubt with the wider political repercussions which the closure of Scotland's sole aircraft manufacturing facility could be expected to cause in the midst of the devolution debate, that on 17th March 1975 Tony Benn, Secretary of State for Industry, announced in Parliament that the Government had decided to alter the criteria governing the scope of public ownership to include companies with a turnover, including that of subsidiaries, of £7½ million instead of over £20 million.

The effect of this revision would have been to bring Fairey Britten-Norman into the new corporation as well as Westland and also Short Brothers of Belfast, in which the Government had a majority shareholding. But Westland continued to be excluded on the grounds that it was a helicopter manufacturer, while the omission of Short Brothers was explained on the potentially contentious grounds that a Northern Ireland organisation could not be part of a British aerospace organisation. The official reason for Fairey Aviation being left out despite a turnover of £13 million was that sales of whole aircraft did not account for the requisite £7.5 million. Yet "whole" aircraft only accounted for a third of Scottish Aviation's total turnover. Therefore when the Secretary of State said simply that the effect of the new parameters "will be to include Scottish Aviation," it was clear that the revision had been specifically aimed to achieve SAL's inclusion in what was then being described as the A.C.G.B., the Aircraft Corporation of Great Britain.

Not surprisingly this development did not in itself allay the fears of those who worked in SAL. Early in May the Prestwick Trade Unions were writing to Tony Benn looking for guarantees that the formation of the new corporation, particularly coming as it did in the course of an international recession in the industry, would not result in the rationalisation and centralisation of certain activities with consequential loss of employment at Prestwick.

On one issue at least Management, Trade Unions and local Members of Parliament were in full agreement and that was the necessity to ensure that the Prestwick factory retained its character and identity within the nationalised corporation. An extensive process of lobbying ensued as the Company pressed its claim to be recognised as the light aircraft specialist in the ACGB. A number of new projects were being evaluated at this time. The most obvious was the development of the Bulldog Series 200, the civil version of which was christened the Bullfinch, essentially a four-seater variant with a retractable undercarriage which would retain the aerobatic capability of the Series 120 but also appeal to the civil market because of its touring capabilities. Although it was recognised that this would be a relatively expensive private aircraft, it was felt that even a modest level of civil sales each year would help to smooth out the "flood and drought" pattern of military procurement.

The Bullfinch, civil version of the 200 Series Bulldog

Detailed consideration was also being given to pursuing the market for agricultural aircraft, specifically the overhaul and spares support of the Leonides engine for this purpose. The Leonides was a 540 horse-power 9-cylinder, super-charged, air-cooled, radial piston engine, designed, developed and originally produced by Alvis during the 1950's and 1960's. Although the engine had established itself in transport and training applications, it was an unknown quantity in the agricultural aircraft role, but Scottish Aviation had assessed it as lighter, quieter and more economic than its competitors and more importantly had recognised that a high proportion even of newly-produced agricultural aircraft used overhauled engines which had been purchased second-hand. SAL had therefore become Alvis' nominated licensee for the engine in this market and was awaiting the results of field trials in Tennessee before deciding whether to pursue the project, although opinions as to its viability were divided from the start.

The Company was also looking to develop its growing reputation in the sphere of military training aircraft. It had already identified trends which indicated a future requirement for a small jet trainer, providing not much more than 1,000 lbs thrust, early in the 1980's. This was the initiative which eventually gave rise to the Company's own design study for the "Sapphire" but in 1975 the immediate line of development which was considered but ultimately rejected was to pursue a suggestion made by Miles that SAL should acquire an option to manufacture its "Student" jet trainer under licence.

Of the various possibilities under examination the project which the Company was most anxious to pursue, not only because it was potentially the most substantial but also because it offered the best prospect of preserving a distinct identity in the new nationalised aircraft corporation was the launch of a civil Jetstream. A number of independent surveys strongly suggested that there was a market, predominantly in North America, for at least 20-25 of these aircraft per annum.

Paradoxically, the very feature which made this prospect so attractive also made the Company wary. Scottish Aviation had cause to remember the effects of pinning all its hopes on a single project in the past and the Twin Pioneer experience lingered in the memory as did the fate

Jetstream production re–commenced at Prestwick, 1973.

of Handley Page. The scale of the civil Jetstream project in relation to the size of the Company as a whole therefore caused it to proceed with caution.
The prolonged uncertainty during 1974 and 1975 as to when the Aircraft and Shipbuilding Industries Bill would actually reach the Statute Book, together with the difficult economic climate at the time, made the Laird Group particularly reluctant to proceed with the project on a private venture basis and plans originally made in 1973 to lay down a batch of ten aircraft for the civil market were shelved early in the following year. However, once it became clear that Scottish Aviation was to be included in the nationalisation of the industry, steps were taken to try to secure Government funding for the civil Jetstream as a long-term export product. Management, Trade Unions and M.P.'s lobbied extensively to secure this goal.

T.Mk. II Jetstreams for the Royal Navy.

In a 200 page submission presented on 31st July 1975 to Lord Beswick, then Minister of State at the Department of Industry but subsequently to become the first Chairman of British Aerospace, Scottish Aviation formally requested Government assistance in two ways. First, it asked that the last 10 of the 26 aircraft being built for the RAF should be released for sale to the civil export market and 10 replacement aircraft concurrently re-ordered by the Ministry of Defence for delivery two years later. The Company's argument was that since the MoD had already deferred the delivery programme and was restricting its utilisation of the aircraft as a result of a lower than expected requirement to train multi-engined pilots, it would be not only consistent with the Ministry's requirements but also in the national interest to make

aircraft promptly available to meet the delivery demands of the civil market, where approximately 25 of the original Handley Page aircraft were already successfully operating both for commuter airlines and corporate organisations.
Secondly, the Company requested Government funding for future production and development of the aircraft for this market, in other words for an investment which would in due course be taken over by the British Aerospace Corporation but which needed to take place without delay to protect a valuable export opportunity.
Neither the Ministry of Defence nor the Department of Industry were prepared to be swayed by these arguments and consequently Scottish Aviation approached the nationalised corporation's Vesting Day on 29th April 1977 with the question of whether it would be able to pursue the Jetstream as a civil project still unresolved, although further evidence of the market for the aircraft in this role came when Century Aviation in the United States made an approach late in 1976 in an attempt to buy the manufacturing rights.

The period between 1974 and 1977 was an uncertain and unsettling one for Scottish Aviation with many of its existing programmes passing their prime and with new work extremely difficult to secure. Not only was the Company's workload dwindling, so also was its most fundamental resource. In 1975 and 1976 reducing workload led to personnel numbers having to be virtually halved by redundancy from some 2,700 to less than 1,400 employees. At the same time the Company lost a number of its most experienced Senior Management including Hugh Laughland who was succeeded as Managing Director by Lyn Phillimore, who had previously been Chief Designer and then Production Director.

An era ends . . . externally at least, December 1977.

On 1st January 1978 Scottish Aviation Limited as a Company passed into history. Although there was an echo of the past in the new designation of the Prestwick factory as British Aerospace, Scottish Division, an era had ended and in the uncertain climate of the time there was an acute perception that it had passed. All those who had been associated with "Scottish" at this time would have preferred to have seen Prestwick enter the new nationalised aerospace corporation on the crest of a wave rather than at a low ebb in its fortunes.

Yet there is obviously more than a mere legal entity about any Company which has managed to survive for over forty years in the particular turbulence of the aircraft industry. In 1945 there were twenty seven distinct airframe design companies in Great Britain. On the eve of the creation of British Aerospace this number had been reduced to half a dozen and Scottish Aviation Limited was one of them. For a relatively small organisation to have achieved this, and to have accomplished so much in the process, is a measure of the qualities of resilience and determination possessed by those who made up Scottish Aviation and, as the subsequent history of Prestwick has shown, these are qualities of the sort which do not simply disappear.

Chapter 10

"ENDINGS and BEGINNINGS" British Aerospace

With the benefit of hindsight it is clear that the inclusion of Scottish Aviation Limited in the new British Aerospace Corporation was not an ending but rather the beginning of a new phase in a continuous history. Left to fend for itself as the recession in the industry plunged to new depths, and standing outside the nationalised body to which the Government was committed, it is hard to imagine that Scottish Aviation could have survived, despite all the tenacity and opportunism which it had displayed in the past.

Yet in the initial years of British Aerospace many at Prestwick felt that they were taking part in a finale rather than an inauguration. It was not so much that people still wanted to think of the place as Scottish Aviation Limited. It was rather a result of the curious fact that, long after Prestwick employees had fallen into the habit of describing themselves as belonging to "British Aerospace, Scottish Division", personnel with whom they came into contact from other parts of the new company continued to refer to them as "Scottish Aviation". Consequently, although it was recognised that being in the new corporation offered a better prospect of security than being outside it, there was a sense of isolation. This arose partially from Prestwick's geographical location some 200 miles from the nearest of the other sites and more perhaps from the fact that it was the only one which had not belonged to any of the three major groups of sites, the former British Aircraft Corporation, Hawker Siddeley Aircraft and Hawker Siddeley Dynamics, which made up the rest of the corporation and which already had well developed inter-relationships. Principally, however, the sense of isolation derived from the continuing uncertainty surrounding the future of Prestwick's work programmes.

This is not to say that steps were not taken to introduce new work into Prestwick to alleviate its situation. During 1978 the Scottish Division undertook various jobs on behalf of other parts of the newly-formed Aircraft Group, including machining for the Kingston factory, work on Tornado ground equipment for Warton and on the Nimrod for Manchester. The latter was followed by the provision of mobile repair party support at RAF Kinloss and then by a package of finishing work on a number of BAe 748 airliners at Prestwick itself.

Last of the 748's completed for B.Ae. Manchester, February 1982.

These programmes, however, were all comparatively small and generally short-term and it was not until the Corporate Board decided, as part of a policy designed to redress the balance between civil and military projects in the industry, to revive the HS 146 short-haul airliner

B.Ae. 146 engine production in The Palace.

that Scottish received a substantial injection of new work which offered a prospect of continuity. The decision to launch the BAe 146 into production was reached in July 1978 by which time design work was already under way with the Prestwick Design Office engaged on the engine pylons, the ailerons and the spoilers. Subsequently production responsibility for the pylons was also allocated to Prestwick. This opportunity to show the Englishmen at Hatfield what the Scots could do was seized with relish, the first production set was completed in February 1981 and Scottish quickly established a reputation for delivering its engine pylons to cost, to quality and to programme.

The project which Prestwick was most anxious to secure, however, and the one which would convince employees that belonging to the new corporation really was the beginning that they hoped it would be, was the chance to launch a new development of the Jetstream. As early as February 1973 the Scottish Aviation Board had considered a paper on a re-engined version of the aircraft to be designated the Series 300. Design studies on the Series 300 had in fact commenced in February 1972 and consideration was given first to the Pratt and Whitney PT6A/41 engine and then to the Astazou XVIII before attention was finally focussed on the Garrett TPE 331, the model from which the Jetstream 31 eventually derived its powerplant.

Not surprisingly it took time to convince the British Aerospace Board to make the substantial investment required to re-launch an aircraft which had failed to capture the market expected of it in its first incarnation and for which the basic design was by now a dozen years old. On the other hand, the failure of the Jetstream under Handley Page had not been the fault of the aircraft itself but of the excessively ambitious plans which had led that Company to over-reach its resources. If there were any criticisms of the 57 Jetstream aircraft in operation before the Series 31, they were centred on the disappointing performance and serviceability of the original Astazou engines, shortcomings which had already led some of the civil operators of the aircraft in America to adopt the Garrett engines which Handley Page had incorporated into the design for the United States Air Force requirement in 1968.

Most important was the fact that civil operators had taken the trouble to make this change and were busy using the aircraft. This in itself drew attention to the change in circumstances which had taken place since the aircraft's original launch. Handley Page had made the mistake not of aiming its product at a market for which it was not suited

- on the contrary the aircraft was extremely well matched to the perceived requirements of the operator - but of producing for a market which had still to emerge. In that sense Handley Page was a victim of its own foresight.

During the decade since the original launch of the HP Jetstream there had been a dramatic growth in regional airline traffic throughout the world as commuting by air became increasingly common and as "deregulation" released large national airlines from the obligation to operate low volume routes thereby spawning a proliferation of small new commuter networks, particularly in North America, Australia and parts of Europe.

Even in the recession of the late 1970's it was evident that this was the fastest growing sector in the aviation business and that the 6-8 seat equipment which had been adequate in the early stages of its development was being superseded, through the growth in demand, by aircraft carrying 15-20 passengers.

Jetstream 31 Interiors in Commuter, Executive Shuttle and Corporate configurations.

In this context the foresight which had gone into the design of the Jetstream gave it a number of outstanding attributes. First, it was unusual for its size in being a pressurised aircraft and therefore capable of cruising, at 25,000 feet, above the variable weather conditions to which passengers would be subject in most of the competitor aircraft then available. Secondly, the aircraft had a unique six foot cabin height, enabling passengers, as both Handley Page and subsequently British Aerospace promotional material pointed out, to "walk tall" in Jetstream whereas they would have to stoop in the more confined space of the competitors' aeroplanes. The importance of these two features lay in the confidence which they imparted both to passengers who were accustomed to travelling in larger aircraft and to those who were not familiar with travelling in aircraft at all. The third vital feature of the Jetstream, again deriving from the fact that it was essentially the product of large airliner thinking scaled down rather than a small general aviation aircraft scaled up, was its extremely durable airframe and long fatigue life. This made it an aircraft which could meet, with low maintenance requirements, the heavy utilisation patterns of the commuter/regional airliner operator.

Another attraction of the Jetstream 31 project to British Aerospace in 1978 was that it was the development of an existing aircraft. The time and costs required to develop it to embody the new Garrett TPE 331-10 engines and Dowty Rotol 4-bladed propellers with their associated system changes could be expected to be relatively modest by comparison with the massive investment which would have been required to initiate an entirely new design. British Aerospace therefore had the prospect of being able quickly to add to its projected "family" of civil aircraft a type which would fit into the fastest growing segment of the market.

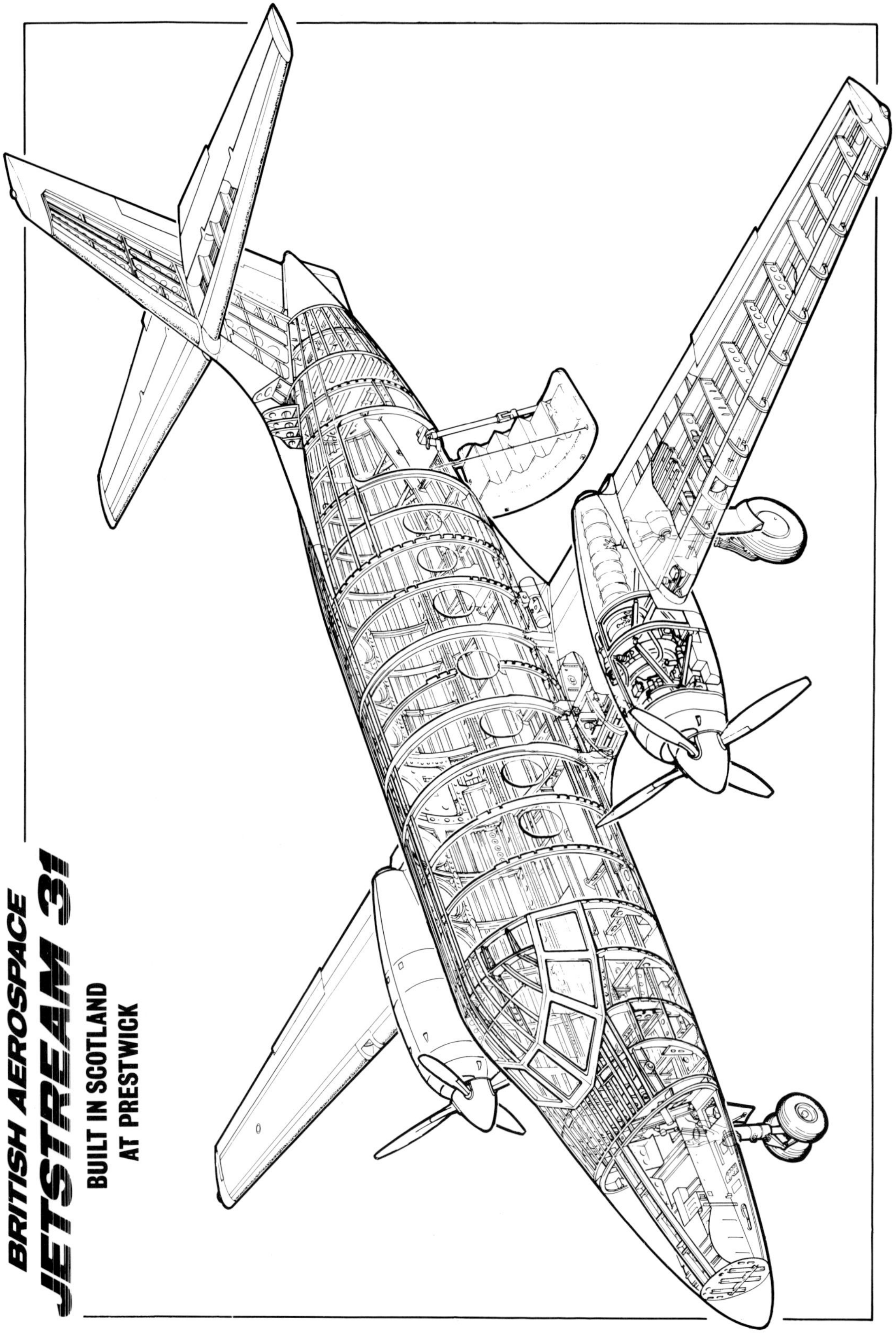
BRITISH AEROSPACE
JETSTREAM 31
BUILT IN SCOTLAND
AT PRESTWICK

Time was certainly of the essence, because the aircraft then competing for this market already had substantial production runs behind them. British Aerospace had to be concerned therefore to determine whether precisely those features which would make the Jetstream technically the best aircraft for the job might not also make it the "Rolls Royce" of the small airliner market in terms of its cost and consequently put it out of reach of the prospective customer.

This problem was compounded by the fact that although the aircraft itself appeared to require a comparatively small investment to develop it into the Series 31, the Prestwick site had suffered badly in terms of lack of new capital investment in its fabric and facilities during the years of prolonged uncertainty between the decision to nationalise the aerospace industry and the implementation of that decision.

It had also to be a source of concern that the organisational infrastructure required to support the Jetstream as a civil aircraft project was significantly different from that which had previously been required to manage the T.MkI and T.Mk.II Series as military aircraft programmes. Whereas military orders tended to be for relatively large numbers of aircraft for a limited number of customers with common specifications, the civil market entailed a much larger number of customers, who, particularly in the case of regional airline and corporate operators, tended to order in small quantities and with a much wider variation in individual specification. It was clear therefore that a full-scale launch of the Jetstream 31 would require a substantial strengthening of a number of functions at Prestwick, including design engineering, sales and technical sales, flight operations and product support.

The decision facing British Aerospace in 1978 as to whether or not it should proceed with the Jetstream 31 project which was being advocated by its Scottish Division was not straightforward. The aircraft itself seemed to be well suited to an emerging market but on the other hand the total resources which would need to be applied to launch the project clearly went well beyond those needed for a simple development of an existing design and extended to a major investment of both funds and personnel to re-configure the Prestwick factory for a significant change in its market orientation.

Not the least of the difficulties with the corporate and small airline market was that, because individual orders were likely to be for no more than either single or perhaps two or three aircraft apiece, it could be expected to take a considerable time to obtain the level of orders which would bring the risk associated with the non-recurring launch costs of the project to an acceptable level.

The development aircraft, G–JSSD, formerly a Series 200.

It is not therefore as paradoxical as it may at first seem that, in pursuit of an order book substantial enough to justify the launch of Jetstream 31 as an essentially civil project, Scottish Division pinned its initial hopes on obtaining a military order for the aircraft. There was an identified Ministry of Defence requirement to replace its ageing communications fleet of Pembrokes, Herons and Devons, a role for which Jetstream seemed ideally suited. Initially it appeared that this requirement might amount to approximately a score of aircraft. The only perceived drawback to what otherwise would obviously have constituted a solid plank in the launch platform was the danger that a single order of this size might prove to be too big in the sense that it would commit virtually the whole of the first two years of production to one military customer and would leave no aircraft available for allocation to civil buyers at a time when Jetstream 31 was already starting from behind its principal competitors and therefore needed to achieve market penetration without delay.

In the event this dilemma was resolved when the MoD requirement was at first deferred and finally cancelled as a result of a Defence Spending Moratorium, but in the meantime it was sufficient to persuade British Aerospace to make an initial commitment to the Jetstream 31 programme and before the end of 1978 approval was given to commence the necessary design, development and pre-production work and of course the marketing effort to secure the customer base on which the decision to launch the aircraft into production would depend.

At this stage it was envisaged that although the Jetstream 31 would eventually become the largest single project at Prestwick and one which would occupy the bulk of the site's capacity, the older projects which had long formed the basis of Prestwick's workload would continue to make important contributions. Although some of them, like the Bulldog, appeared to be approaching the end of their product life cycle, others seemed to be evergreen.

Unfortunately external circumstances were to impinge dramatically on this situation. By 1978 Prestwick had achieved the rare distinction of successfully retaining the contract to overhaul and repair the aircraft of the Canadian Air Force's wings in Europe through 23 successive annual negotiations. Since 1955 Scottish had handled a series of different aircraft types from the Sabre to the CF.104, responding to the new levels of sophistication required with each change. In 1978, however, a new factor came into the equation as unemployment in Canada rose to the then unprecedented level of 10%. In these circumstances the allocation of the Canadian Air Force's servicing contract to a "foreign" contractor acquired a new political dimension. Suddenly the familiar roar of jet fighters disappeared from Prestwick, the chains of the runway arrestor gear lay rusting and unused and the repair and overhaul of the CAF units deployed in Europe was assigned to North West Industries on the other side of the Atlantic.

"Run–down of the run–up": dismantling the CF104 engine bed.

The Bulldog was also to some extent an indirect victim of the recession. Being a piston-engined aircraft its operating costs were highly sensitive to the availability and consequently the price of Avgas fuel which had become radically more expensive in the aftermath of the oil crisis in 1973. Although the Bulldog itself had sold well against other piston-engined types in the very competitive market for primary trainers, it and its proposed derivative the Bullfinch now found themselves up against designs for small jet trainers, and in particular turbo-prop aircraft. The turbo-prop engine, while more expensive in the first instance, was generally more reliable, had longer life between overhauls, was lighter in weight and used Avtur which was only about half the price of the fuel required by the piston- engine. After 1978 there were very few new orders for Bulldog.

Jordanian Bulldogs departing by Hercules, April 1981.

Essentially the same pressures curtailed the Company's engine overhaul activities at Prestwick. From the autumn of 1979 the department which had previously been Scottish Air Engine Services saw its workload diminish rapidly and worldwide efforts to find more work on the old Pratt and Whitney piston engines proved unavailing. In October 1980 it was decided to close the shop and merge its residual MoD work on Lycoming engines, the Bulldog's powerplant, with the Rolls Royce Griffon line. But sadly engine overhaul went down to a double punch and the second blow fell even before the decision to establish an integrated overhaul facility had been fully implemented. In April 1981 the Ministry of Defence finally announced its intention to phase the Shackleton out of service and to replace it with the Nimrod. The effect on Scottish Division was to terminate the requirement for the Griffon engines which had long been the backbone of its Rolls Royce engine contracts. At the peak of the programme Prestwick had been overhauling 240 engines per annum. In the autumn of 1981 Prestwick's Griffons became extinct.

The last Pratt and Whitney engine to be overhauled, April 1981.

The last Rolls Royce Griffon Engine to be overhauled, May 1981.

Fortunately, despite the absence of a Ministry of Defence order for the aircraft and despite the other evident risks, the Corporate Board of the newly de-nationalised British Aerospace, now P.L.C., had given its approval in January 1981 for the full production launch of the Jetstream 31 with a view to reaching a mature production rate of 25 aircraft per annum. At the same time a new Managing Director, Brian Thomas, formerly Production Director of the Weybridge-Bristol Division, was appointed. Prestwick began to pulse with a new sense of urgency. The Jetstream's full production launch provided the opportunity to redeploy the labour shed by the defunct engine overhaul programmes and also to ride the next storm to hit the Division when, early in 1982, Lockheed's decision to cancel its TriStar airliner programme led to the termination of Prestwick's sub-contract work on that aircraft and, more importantly, to the withdrawal to Lockheed's own factories in the United States of the C.130 fuselage panel and pylon work which Scottish had performed for the past 16 years, producing at one stage as many as seven aircraft sets per month.

The final set of C.130 panels to be manufactured at Prestwick, December 1982.

The last set of Hercules panels left Prestwick on 17th December 1982 and Prestwick went into the new year as a very different organisation from the one which it had been five years earlier. It had changed from being essentially a sub-contractor engaged on a number of comparatively small but long-running programmes in a predominantly military aircraft environment to being, apart from its responsibility to supply 146 engine pylons to Hatfield, a prime contractor in the civil market with a single product which was more sophisticated and demanding than any project it had undertaken before.

Clearly the infusion of the new personnel with prior experience in this business from other parts of British

Aerospace contributed to the speed with which Prestwick adapted to its new role but the response to the challenge posed by this change came from throughout the site.

The prototype aircraft, G-JSSD, a former Series 200 which had been obtained in the United States, re-engined and re-worked to carry out the bulk of the development flying associated with the programme, made a first flight of 1 hour 22 minutes duration on 28th March 1980. By the time that the first production aircraft, the aptly registered G-TALL, was presented to the world in 1982 in a dramatic roll-out ceremony timed to coincide on 25th January with the birthday of another of Ayrshire's famous sons, Robert Burns, the development programme had logged 440 flights and 550 hours of flying including tropical trials of the aircraft in hot and high conditions in Spain. G-TALL joined the development programme after making its first flight on 18th March and the reward of all the endeavour required to support this process duly came when the UK type Certificate of Airworthiness was awarded on 29th July and the US Certificate of Airworthiness on 1st December 1982.

The Roll Out of the first production Jetstream 31, 25th January, 1982.

A hectic year, which had begun with the roll-out of the first production aircraft, culminated in two more ceremonies. On 15th December the first customer delivery was made to Contactair of Stuttgart, one of the leading air charter operators in West Germany. Fifteen days later the first aircraft to a UK customer was handed over to a Scottish operator, Peregrine Air Services of Aberdeen, to be used on oil-related charter work. Significantly both Contactair and Peregrine quickly translated a contract option into a firm order for a second aircraft.

First customer : Contactair aircraft over the Alps.

The opening of the new final assembly and equipping line, B.G. Thomas and the Secretary of State for Scotland, the Rt. Hon. George Younger, M.P.

However, by the end of 1982 firm sales and contract options only totalled 14 aircraft and with the exception of one demonstrator there had been no deliveries to the United States, which had been expected to represent the predominant demand for the aircraft. Unfortunately the civil aircraft market as a whole was remaining in a state of recession for longer than had been anticipated. In an effort to stimulate sales Scottish Division embarked on a world-wide series of Jetstream 31 demonstration tours.
Until such time as production aircraft were available for demonstration purposes, and in order to bring Jetstream 31 quickly to the attention of prospective customers, the Division had re-furbished an old Series 200 fuselage as a mock-up. 47 feet long and some 15 feet high on its trailer, this had made its first public appearance at the Paris Air Show in 1981 and, configured in an 18-seat commuter airline layout with a fully representative cockpit, it went on to exhibit the aircraft's superior cabin accommodation and facilities at the Cranfield, Hanover and Farnborough Air Shows and to go on tour throughout Europe and the United States as well as making a more light-hearted appearance in the Lord Mayor's Show in London in 1982 when it was suitably attired in a large tartan "tammy" and sash.

The fuselage Mock–Up in the Lord Mayor's Show, London 1982.

By the summer of 1982 production aircraft were available for demonstration purposes and a series of four tours starting in August of that year and finishing in the following May progressively covered Western Europe and Scandinavia, the United States and Canada, Southern Europe and North Africa and finally a marathon 7-week, 34,000 mile tour of the Middle East, South-East Asia and Australasia, a series of gruelling schedules in which the Jetstream 31 fully lived up to its reputation for ease of maintenance and high despatch reliability.

In the same month as the fourth tour was completed Jetstream 31 went into service with its first American operator, Atlantis Airlines based in Florence, South Carolina, who had ordered two aircraft and inaugurated their services on 15th May 1983 on the Charlotte-Knoxville route.

The demonstration tours and the delivery of aircraft into operation quickly proved that there is no substitute for showing off the hardware. The first half of 1983 was quiet in terms of sales with an order from the newly-formed Birmingham Executive Airways in March to fly routes, previously operated by British Airways, linking Birmingham with Copenhagen, Milan and Zurich, followed by one from the long-established, prestigious charter operator, McAlpine Aviation, based at Luton. Following the pattern of earlier operators both soon returned for additional aircraft. The presence of aircraft out and about in the market-place was bearing fruit.

On 28th August the Company was able to announce a highly important breathrough into the Australian market where Skywest Airlines, the country's largest third-level operator, had placed an order for three aircraft to link Perth with a score of other cities and towns in Western Australia. A particular requirement of this order was the installation of a water methanol injection system in the aircraft's engines to compensate for the effect on performance of the extreme temperatures experienced during the summer. At the time this development, entailing as it did "hot and high" trials in the Nevada desert and in California, represented simply another hurdle to be crossed in the already onerous process of the project's launch and customer acceptance phase, but it was one which was to stand Jetstream in good stead for westward sales penetration in the USA.

At the end of November the Jetstream secured its biggest order to date when Vee Neal Airlines of Latrobe, Pennsylvania, confirmed its earlier selection of six aircraft to operate routes in the North-Eastern United States and Canada. It was not only the size of this particular order which suggested that the aircraft was succeeding in securing customer acceptance. It was perhaps the sincerest form of flattery that at the same time as it confirmed its order the airline announced that from 1st December it was changing its name to Jetstream International Airlines. The first two aircraft were handed over at Erie three days before Christmas.

Jetstream International's position as the world's largest operator of the aircraft did not last long, because 1984's sales activity simply picked up where 1983 had left off. On 16th January Metro Airlines of Houston, Texas, one of the most highly-regarded commuter airlines in the United States, announced that it planned to acquire twelve aircraft to be used in a new subsidiary operation called "Eastern Metro Express".

Consequently when British Aerospace announced in the middle of this intense sales activity in January 1984 that it was re-organising the operating units in its Aircraft Group into three new Divisions, rather than the six into which it had originally been formed, Prestwick could take pride in the fact that, while it was losing its "Scottish Division" title, it was demonstrably going from strength to strength and could take its place along with the Hatfield, Filton and Chester factories as a vital part of the new Civil Division.

Demand for Jetstream 31 had reached a pitch where, in April 1984 and even before the project had reached its previously approved mature rate of production of 27 aircraft per annum, authorisation was given for production to be escalated to a new annual rate of 36.

By the end of 1984 a surge of orders had taken sales of the aircraft to 69. In 1984 alone 44 aircraft had been sold, 37 of them for export and 29 into the intensely competitive American market. New customers included the Royal Navy with an order for four aircraft, designated Jetstream T.Mk.III, to be used as observer trainers, Swedair, Sweden's largest regional airline, and a new Dutch operator, NetherLines, which ordered four aircraft for scheduled services within Holland and to Luxembourg, Germany and the United Kingdom.

Perhaps the most significant feature of 1984's sales, however, was the emergence of a new scale of commuter airline operation in the United States. The Eastern Metro Express deal paved the way, the name of this new venture reflecting the collaboration between the commuter airline, Metro, and Eastern Airlines, one of the major national carriers. Adopting Eastern's livery and more importantly its ticketing facilities, EME operated on a hub-and-spoke

Jetstream 31 operators.

BIRMINGHAM EXECUTIVE

ROYAL NAVY

EASTERN
EASTERN ATLANTIS EXPRESS
N157AA

31-620
Skywest

Republic Express
Republic

Britanni

basis between major airport and outlying regions, thereby giving the air traveller a new integrated service.
The pattern was repeated with another new Metro affiliate, American Eagle, which announced its selection of 10 Jetstream 31 aircraft in November 1984 to feed the American Airlines hub at Dallas/Fort Worth, Texas, rapidly becoming the busiest airport in the world, from 15 cities in the South-Western United States. Simultaneously one of the original American customers revealed that it had entered into a similar arrangement when Atlantis Airlines ordered a further six aircraft for an operation named Eastern Atlantis Express to serve Eastern's hub at Charlotte in North Carolina as well as Atlanta, Georgia. Atlanta was also the source of the biggest sale to date, when in March 1985 Phoenix Airline Services based there ordered twenty aircraft to link a similar number of cities with Republic's hub at Memphis in an operation to be known as Republic Express Airlines.

"Hub–and–Spoke" commuting in operation.

In a short space of time the Jetstream 31 had established itself firmly in the market. It had done so for reasons which bore out the quality of the original design and the requirements demanded of it. The "Walk Tall" cabin and large airliner "feel" of Jetstream gave it high customer acceptability. During its first year of operations Eastern Metro Express, for instance, reported a 400% growth in traffic, carrying 7000 passengers in its first month of scheduled service and 28000 in the last month of the year. By that time the airline was averaging 2350 departures a month and this intensity was by no means untypical of other Jetstream 31 operators many of whom were by now logging 2500 to 3000 flying hours or more per annum. At these levels of utilisation, five or six times greater than that of the typical corporate aircraft, the Jetstream 31's rugged airframe and its long maintenance life came into their own. During Jetstream's first year of operation the aircraft achieved a despatch reliability in excess of 99%.

Jetstream 31 final assembly in the "Britannia" Hangar.

So far as Prestwick was concerned the most obvious result of the aircraft's evident suitability for hub-and-spoke operations was that individual orders were now frequently very much larger than had originally been anticipated when the project was launched. This also meant that within a few months of authorising one production rate increase the British Aerospace Board was being asked to give its approval for another, this time to a mature rate of 48 per annum, and when this was announced on 5th September 1985 the distinction which the Jetstream 31 and its producers had achieved was recognised by a visit to the Prestwick site by the Right Honourable Margaret Thatcher, MP, the first-ever visit to the site by a British Prime Minister.

By the end of the year sales of the aircraft had soared past the hundred mark, with 51 aircraft sold in 1985 alone.

The Queen's Award for Export Achievement, 9th August 1985.

The Prime Minister's Visit, 5th September 1985: The Rt. Hon. Margaret Thatcher, M.P. with Netherlines air crew and (1 to r) the Secretary of State for Scotland, G.J. Curran, General Manager, Prestwick and J.L. Glasscock - Managing Director, Civil Aircraft Division.

Prestwick's 50th anniversary jubilee year in 1985 was indeed golden, in marked contrast to the depression and uncertainty which had pervaded the site's activities on its entry into British Aerospace eight years earlier. The transition was underlined by a very welcome birthday present. On 9th August 1985, the anniversary of the foundation of the predecessor company, British Aerospace Prestwick received from the hands of Her Majesty's Lord Lieutenant for Ayrshire and Bute, the Queen's Award for Export Achievement in recognition of the £41 million of overseas sales achieved by the Jetstream during the previous year. It was a gift presented at a ceremony for the whole workforce and one in which all could justifiably share.

It is appropriate too that this account of the history of Scottish Aviation should close on 9th August 1985, fifty years after it had begun. "SAL" as a company had of course gone by then, but so for that matter had "British Aerospace, Scottish Division". Prestwick was now part of the re-named Civil Aircraft Division in the reorganised British Aerospace.

But the evening of 9th August belonged not to Scottish Aviation, nor Scottish Division, nor Civil Aircraft Division, Prestwick, but to their employees, to those who had made each of these organisations in turn and had given them a continuity and an existence beyond mere names. They came that night to an Anniversary Dance, employees present and former, young and old with their partners, nearly 3000 in all. It was an occasion of which David McIntyre would have approved, celebrated in a way which would have delighted him, in a hangar which had been extended to twice its normal length by the addition of a huge marquee.

" The evening belongs to the employees "

Walking tall, flying high.

The evening was haunted by the spirits of Prestwick past. The hangar resounded to the music of the big band era, conjuring up memories of the American aircrews, the Liberators and Flying Fortresses for whom Prestwick had once been a second home. Beyond that the distant drone of aero engines announced the arrival of other hardy sons, aircraft associated with the Company returning for the Anniversary Open Day. The next morning revealed them as it had revealed the overnight arrivals of the Transatlantic air ferry 40 years before: Tiger Moth, Anson, Mosquito, Dakota, CF.104, Twin Pioneer, Bulldog, Jetstream. Makers of history, each in their own way, they more than anything represented the continuous process of change which the Company had gone through during fifty years and was going through still.

For that reason perhaps the lasting impression of the evening of 9th August 1985 was not one of retrospection but of an energy that was of the present and directed towards the future. Amidst all the other changes of half a century one thing at least remains unaltered at Prestwick, a collective belief that there is something special and important about building and selling aeroplanes, and a determination to ensure that the Scottish lion continues to be both rampant and winged.